MULTIPLE MODERNITIES AND GOOD GOVERNANCE

This book represents the first discussion from a political science perspective of the concept of Multiple Modernities in three dimensions. First taking stock of the discussions of the concept itself, the book then connects the concept to more recently developed analytical and normative concepts that concretize it, before finally opening up a discussion about its implications and consequences for the political dimension.

Written by outstanding scholars in the field, the book addresses four principal concepts – Good Society, Good Governance, Human Security and Varieties of Capitalism. It determines whether and to what degree these concepts enable us to discover the commonalities and differences that distinguish the emerging multiple modernities in our time with respect to their political implications and consequences.

This text will be of key interest to scholars and students of political theory, political economy, international relations, comparative politics and sociology.

Thomas Meyer is Emeritus Professor of Political Science at the Technical University of Dortmund, Germany and Editor-in-Chief of the monthly political magazine, *Neue Gesellschaft/Frankfurter Hefte Journal of Social Democracy*.

José Luís de Sales Marques is President of the Board of Directors of the Institute of European Studies of Macau.

GLOBALISATION, EUROPE, MULTILATERALISM SERIES

This series delves into a given dynamic shaping either the global-regional nexus or the role of the EU therein. It offers original insights into globalisation and its associated governance challenges; the changing forms of multilateral cooperation and the role of transnational networks; the impact of new global powers and the corollary multipolar order; the lessons born from comparative regionalism and interregional partnerships; as well as the distinctive instruments the EU mobilises in its foreign policies and external relations.

Series Editor: *Mario Telò, Institut d'Études Européennes at the Université Libre de Bruxelles, Belgium.*

Series Managed by: *Frederik Ponjaert, Université Libre de Bruxelles, Belgium.*

For more information about this series please visit: www.routledge.com/Globalisation-Europe-Multilateralism-series/book-series/ASHSER1392

The Politics of Transatlantic Trade Negotiations
TTIP in a Globalized World
By Jean-Frédéric Morin, Tereza Novotná, Frederik Ponjaert and Mario Telò

Interregionalism and the European Union
A Post-Revisionist Approach to Europe's Place in a Changing World
By Mario Telò, Louise Fawcett and Frederik Ponjaert

Developing EU-Japan Relations in a Changing Regional Context
A Focus on Security, Law and Policies
Edited by Dimitri Vanoverbeke, Takao Suami, Takako Ueta, Nicholas Peeters and Frederik Ponjaert

Deepening the EU-China Partnership
Bridging Institutional and Ideational Differences in an Unstable World
Edited by Mario Telò, Ding Chun and Zhang Xiaotong

Multiple Modernities and Good Governance
Edited by Thomas Meyer and José Luís de Sales Marques

MULTIPLE MODERNITIES AND GOOD GOVERNANCE

Edited by Thomas Meyer and José Luís de Sales Marques

LONDON AND NEW YORK

First published 2018
by Routledge
2 Park Square, Milton Park, Abingdon, Oxon OX14 4RN

and by Routledge
711 Third Avenue, New York, NY 10017

Routledge is an imprint of the Taylor & Francis Group, an informa business

British Library Cataloguing-in-Publication Data
A catalogue record for this book is available from the British Library

Library of Congress Cataloging-in-Publication Data
Names: Meyer, Thomas, 1943- editor. | Sales Marques, Jose Luis de, editor.
Title: Multiple modernities and good governance / edited by Thomas Meyer & José Luís de Sales Marques.
Description: New York : Routledge, [2018] | Series: Globalisation, Europe, multilateralism series | Includes bibliographical references and index.
Identifiers: LCCN 2017059108 | ISBN 9781138574526 (hardback) | ISBN 9781138574533 (pbk.) | ISBN 9781351273886 (ebook)
Subjects: LCSH: Political sociology. | Comparative government. | Globalization--Political aspects. | Human security. | Capitalism--Political aspects. | Civilization, Modern.
Classification: LCC JA76 .M766 2018 | DDC 306.2--dc23
LC record available at https://lccn.loc.gov/2017059108

ISBN: 978-1-138-57452-6 (hbk)
ISBN: 978-1-138-57453-3 (pbk)
ISBN: 978-1-351-27388-6 (ebk)

Typeset in Bembo
by Taylor & Francis Books

CONTENTS

ILLUSTRATIONS

Figures

Tables

Boxes

CONTRIBUTORS

Amitav Acharya is Distinguished Professor of International Relations and the UNESCO Chair in Transnational Challenges and Governance at the School of International Service, American University, Washington, D.C. His recent books include *The End of American World Order* (Polity, 2014; Oxford India, 2015); *Rethinking Power, Institutions and Ideas in World Politics: Whose IR?* (Routledge 2014); and an edited volume, *Why Govern: Rethinking demand and progress in global governance* (Cambridge, 2016). His articles have appeared in *International Organization, International Security, International Studies Quarterly, Journal of Asian Studies, Journal of Peace Research,* and *World Politics.* Professor Acharya is a past president of the International Studies Association (2014–15). During the 2015–16, academic year, he served as the Distinguished Visiting Professor in the Schwarzman Scholars Programs of Tsinghua University, Beijing.

Nassir Abdulaziz Al-Nasser assumed his position as the High Representative for the United Nations Alliance of Civilizations on March 1, 2013. Throughout his career, he has contributed to advancing the multilateral agenda in the realms of peace and security, sustainable development, and South-South cooperation. Previously he held the position of President of the 66th Session of the United Nations General Assembly from 2011 to 2012. Mr. Al-Nasser's career as a diplomat spans more than three decades. Since 1998, he has represented Qatar in the United Nations and has worked on core issues such as security, terrorism, poverty, hunger, and natural disasters. He has received numerous awards and honorary doctorates from various countries and universities for his work on fostering cross-cultural understanding.

Manuel Castells is Professor Emeritus of Sociology and of City & Regional Planning at the University of California, Berkeley. He is also Director of Research, Department of Sociology, University of Cambridge, and a fellow of St. John's College, University of Cambridge. He previously served as Distinguished Visiting Professor at MIT and Oxford University. Professor Castells is the author of 30 books, including the trilogy, *The Information Age: Economy, society, and culture* (Blackwell, 1996–2003), which has been translated into 22 languages, including Chinese. His latest publications are *Communication Power* (Oxford, 2009) and *Networks of Outrage and Hope: Social movements in the internet age* (Polity Press, 2015). He has received, among other distinctions, the Holberg Prize and the Balzan Prize.

Rodney Bruce Hall is Professor of International Relations at the University of Macau. He previously taught at Brown University, the University of Iowa and the University of Oxford. His books include *Central Banking as Global Governance: Constructing financial credibility* (Cambridge, 2008), *National Collective Identity: Social constructs and international systems* (Columbia, 1999), *The Emergence of Private Authority in Global Governance* (Cambridge, 2002), and *Reducing Armed Violence with NGO Governance* (Routledge, 2015). Professor Hall's articles have appeared in *Global Society, International Organization, International Political Sociology, International Relations, International Studies Quarterly, Journal of International Relations and Development,* and *Security Studies.*

Inge Kaul is adjunct professor at the Hertie School of Governance in Berlin and former Director of the Offices of the Human Development Report and Development Studies at the United Nations Development Programme (UNDP) in New York. She has published widely on issues of global governance and international cooperation. Professor Kaul was the lead editor of *Providing Global Public Goods* (Oxford, 2003) and *Managing Globalization and the New Public Finance: Responding to global challenges* (Oxford, 2006). She co-authored the *Governance Report 2013* (Oxford, 2013) and was the editor and co-author of *Global Public Goods* (EE, 2016). Her current research concerns global public economics and finance.

Jürgen Kocka has taught history at the University of Bielefeld (1973–88), the Free University of Berlin (1988–2009), and, as a Visiting Professor, at UCLA (2009–15). He was President of the Social Science Center Berlin (WBZ) and a permanent fellow of both the Berlin Institute of Advanced Study (Wissenschaftskolleg) and the Center for Work and the Human Lifecycle in Global History at Berlin's Humboldt University. He holds honorary degrees from several European universities and is the recipient of many distinguished prizes, among them the International Holberg Memorial Prize of 2011. Professor Kocka's scholarship focuses on the social, economic, and cultural history of Germany and Europe, comparative history from the 18th century to the present, and patterns of cooperation between the discipline of history and the social sciences. His publications in English include *Industrial Culture and Bourgeois Society: Business, labor, and bureaucracy in modern Germany* (1999), *Civil Society and Dictatorship in Modern German History* (2010), and *Capitalism: A short history* (2016).

Thomas Meyer is Professor Emeritus of Political Science at the Technical University of Dortmund and Editor-in-Chief of the monthly political magazine, *Neue Gesellschaft/Frankfurter Hefte* and the quarterly *Journal of Social Democracy*. He has held visiting professor and guest lecturer positions at numerous universities, particularly in East and Southeast Asia, including Todai University, Tokyo; Beida University, Beijing; and the Indian Institute of Management, Bangalore. He directed projects for the prestigious German Research Foundation (DFG) on such topics as political communication in the media (1995–2001) and the theory and practice of social democracy (2002–06). From 2000–07 he served as Academic Advisor to the European Commission for the Social Sciences and Humanities, and from 2004–08 he was a member of the GARNET Network of Excellence, funded by the European Commission. Professor Meyer's research interests include comparative social democracy, European studies, mass media and politics, religious and political fundamentalism, and the cultural foundations of politics. Among his many books are *The Theory of Social Democracy* (Polity Press, 2007), *Identity Mania* (Mosaic Books, 2001), and *Media Democracy* (Polity Press, 2002).

Julian Nida-Rümelin is professor of philosophy and political theory at the Ludwig-Maximlians-University in Munich. He previously taught at Minnesota State University, the University of Tübingen, the University of Göttingen, Humboldt University Berlin, and the University of Munich. Professor Nida-Rümelin was elected President of the German Association for Analytical Philosophy (1994–97) and President of the German Society for Philosophy (2009–11). He is a member of several academies, including the Academy of Sciences and Humanities, Berlin, and the European Academy of Sciences and Arts. For several years he took a hiatus from academia, serving as a state minister for culture and media in the first national government led by Chancellor Gerhard Schröder (1998–2002).

Tak-Wing Ngo is Professor of Political Science at the University of Macau. He specializes in state–market relations, regulatory governance, and the political economy of development in East Asia. Professor Ngo, who holds a PhD from SOAS (London), worked as an anti-corruption official and journalist before joining academia. He taught at Leiden University and occupied the IIAS Chair in Asian History at Erasmus University Rotterdam. He is the editor of the refereed journal *China Information* and of two book series, *Governance in Asia* (NIAS Press) and *Global Asia* (Amsterdam University Press).

José Luís de Sales Marquez has been President of the Board of Directors of the Institute of European Studies, Macau (IEEM) since January of 2002. From 1993–2001, he served as the Mayor of Macau. His research, teaching, and writing focus on Asia–Europe relations, EU–China dialogue, regional integration, and urban studies.

Jack Snyder is the Robert and Renée Belfer Professor of International Relations in the Department of Political Science and the Saltzman Institute of War and Peace Studies at Columbia University. His books include *Religion and International Relations Theory* (Columbia, 2011), *Electing to Fight: Why emerging democracies go to war* (MIT, 2005), co-authored with Edward D. Mansfield, and *Myths of Empire: Domestic politics and international ambition* (Cornell, 1991).

Julia Tao teaches philosophy at City University of Hong Kong. Her teaching and research interests are focused on moral and political philosophy, public policy and democratic governance, and comparative East–West ethics. She has published in the *Journal of Applied Philosophy, Journal of Social Philosophy, Journal of Medicine and Philosophy, Journal of Chinese Philosophy*, and *Journal of Philosophy East and West.*

Mario Telò is the Jean Monnet Chair of International Relations at the Université Libre de Bruxelles, where he coordinates the Global Europe Multilateralism (GEM) international doctoral program and was past president of the Institute for European Studies. He also teaches at the LUISS University and School of Government in Rome, and has been a visiting professor at numerous universities worldwide. Professor Telò has served as a consultant to the European Council, European Parliament, and European Commission. He is the author of 32 books and over 100 scientific articles, published in seven languages. His recent works include *Europe: A civilian power?* (Palgrave, 2006), *European Union and the New Regionalism* (Routledge, 2014), *International Relations: A European perspective* (Routledge, 2016), and *Regionalism in Hard Times* (Routledge, 2016). He actively participates in public debates concerning international relations and the future of the European Union.

ACKNOWLEDGMENTS

This edited book is the main output of a research conference held at the Institute of European Studies Macau (IEEM) in November 2016 at the invitation of the President of the Institute José Sales Marques and Thomas Meyer, Visiting Professor of the Macau Institute and former Chair of the department of Political Science at the Technical University Dortmund, Germany.

The Conference was conducted jointly by both aforementioned persons and Prof. Mario Telò, Université Libre de Bruxelles.

The editors wish to thank Beatrice Lam of the IEEM for her excellent assistance regarding the Conference. The papers have been redrafted by the authors taking into account the discussions and the contributions of many colleagues of the Macau university, journalists, and IEEM students.

Our special thanks go to Professors Lew and Sandy Hinchman of the USA, who provided the editors with their professional and passionate assistance in all aspects of the editorial work.

FOREWORD

We live in a world characterized by tremendous, permanent, and often unexpected changes, whether in Europe, Asia, the Americas, or elsewhere around the globe. While some of these changes may stimulate social progress, others carry new and unprecedented risks touching every sphere of life: economic, political, cultural, ecological, and religious. Some even threaten the basis for peaceful cooperation within and among societies. They may unleash civil wars or foment acts of terrorism carried out in the name of cultural identity.

People from countries all over the globe greeted modernity—including the culture of cosmopolitanism with which it is associated—as a great hope for permanent progress and peace. But now prospects have grown less sanguine due to a series of economic and cultural crises that have arisen in the arch-modernizing West, from the mistreatment of nature to the destructive power of an untamed market economy that threatens social cohesion and security. The worldwide credibility of the West is at stake, along with its universalistic aspirations. Non-Western actors, including some neo-authoritarian regimes seeking legitimacy through traditionalist cultural identity politics, challenge the West and its liberal world order, both normatively and in terms of power politics.

During this phase of global re-orientation, the cultural factor – the existence of diverse world-views and identities – plays a new and highly ambivalent, even contradictory, role in practical politics, ideological debate, and intellectual discourse. To be sure, there is an increasing readiness to engage in dialogue to found a new global ethic and better forms of mutual understanding and activity. Yet both the trend to defy universalistic norms and values in the name of cultural regionalism and the trend toward aggressive religious fundamentalism in the form of identity politics have become much more forceful and conspicuous. Some economically successful authoritarian models of development in East Asia now seriously rival

Western modernity, while Islamic State activities in Syria, Iraq, and beyond exert tremendous attraction on young people who feel alienated in their own societies.

Given these realities, the Institute of European Studies of Macau (IEEM) sought to promote debate on the issue of modernity by sponsoring a conference in November of 2016. Located in a region marked by many centuries of multicultural experience and co-existence, the city of Macau understands itself as a bridge between East and West. The present volume emerged from the contributions of the prominent thinkers from around the world who attended the gathering. It addresses a wide range of questions. In view of cultural regionalism, are there multiple ways to be modern? To what degree are the West's claims to universality justified? What are the characteristics of the diverse models of modernity and what, if anything, do they have in common? Are there schemes for practical reform that can revive the credibility of the Western model and at the same time be shared by its still-modernizing rivals? How can we more effectively organize intercultural dialogue under conditions of rapid social change? What are the most promising and feasible ways to pre-empt deep social and economic conflicts of transcultural relevance? Are there concepts, values, or ethical and political ideals that bolster intercultural cooperation and recognition because they can be accepted by all? In short, does a strong basis for cooperation and understanding among the principal global actors and adversaries exist, and – if so – what are its limits?

In the framework of the UN, in academic discourse, and in a broad variety of political associations and fora, three concepts have received particular emphasis in recent decades, all raising the implicit claim to have provided (partial) answers to these questions: namely, good society, good governance, and human security. Each lays claim to advancing a potentially cross-cultural vision, one that promotes a form of unity in diversity that is acceptable to all. The overriding objective of the conference that inspired this volume was to examine whether and to what degree such concepts might enable us to discover the commonalities and differences among emerging forms of modernity. In light of the political challenges of the present day, few projects appear more urgent or compelling.

José Luís de Sales Marques, President
Institute of European Studies of Macau

INTRODUCTION

Thomas Meyer

The present volume represents a step forward in the discussion of the concept of multiple modernities in three respects: It takes stock of the discussions about the concept itself; it connects multiple modernities to more recently developed analytical and normative concepts that concretize it; and it attempts to initiate a discussion about the concept's political implications and consequences, particularly by testing its analytical value in respect to key issues concerning political economy and the cultural claims of certain authoritarian regimes, such as China. It is one of the few contributions to the relevant discussion from a political science perspective that draws on the insights of outstanding scholars in the field.

Across the globe, various actors and cultural-political tendencies are increasingly successful at challenging the Western model of modernization and its philosophical underpinnings. They maintain that the present crisis offers final proof of the inherent contradictions or even the bankruptcy of the Western way of life in every sphere, particularly for the establishment of just and inclusive government. They further claim that the shortcomings and crises of Western modernity show that the world needs fundamentally different models of development, culture, and government. Among the new actors that have challenged the culture of modernity as understood in the Western world and offered wide-ranging alternatives to it, two deserve special attention. First, there are various forms of political authoritarianism that emphasize and draw strength from the particularity of national or regional cultural traditions. Second, there are several variants of religious and political fundamentalism that pursue the politics of cultural identity and advocate some type of theocracy.

In particular, during this phase of global re-orientation, the cultural factor – i.e., different cultural world-views and identities – plays a new and highly ambivalent, often contradictory role in politics, ideological debate, and intellectual discourse. To be sure, we are witnessing an increasing readiness to engage in dialogue aimed at founding a new global ethics and better forms of mutual understanding and interaction. Yet both the trend to defy universalistic norms and values in the name of cultural regionalism and the trend towards aggressive religious fundamentalism in the form of identity politics that attacks the very foundations of human civilization have become much more conspicuous and forceful. Some economically successful authoritarian models of development in East Asia, including those of China, Japan, and Singapore, have become serious rivals to Western modernity. Meanwhile, Islamic State (IS) activities in Syria, Iraq, and increasingly beyond exert tremendous attraction on young people who are alienated in their own societies in Europe and parts of Asia and Africa. The recent victories over IS in those two countries may simply displace the group's "caliphate" into other countries, such as Libya or Mali.

The European Union understands itself as a culturally pluralist, democratic polity and a proactive actor for peaceful cooperation, conflict resolution, and development in the world as a whole. Cultural diversity can be an asset that supports a society's creativity and wealth production as well as its cultural productivity and richness. But it can also become a great risk for everybody when emergent socio-economic conflicts – especially in times of crisis – are interpreted in terms of cultural identity politics by entrepreneurs from various cultural, political, and religious communities. The latter seek to profit from instigating intercultural confrontation and hostility.

At this juncture, the multiple modernities approach suggests itself as an analytic tool and a practical guide to tackle the new situation. It was initially conceived by the renowned Israeli sociologist Shmuel N. Eisenstadt in the 1970s who wanted to challenge the hitherto undisputed monopoly held by the notion of Western values and institutions as the universal form of modern culture and politics. The latter position had been dominant in the discourse of modernity initiated and shaped by academic and political debates in the West since the onset of the post-colonial era following the end of World War II. The paradigmatic theories from this era were developed by American scholars such as political scientists Walt Rostow (1960) and Daniel Lerner (1958). While recognizing that the process of modernization would proceed at different rates of speed across nations and cultures, these scholars still considered it to be a process that ultimately would culminate in the emergence of Western-style modernity everywhere. Their uniform model, combining both analytic and normative elements, included the most crucial political, cultural, and economic institutions. Remaining religious differences would not matter much, because religion gradually would succumb to secularization, another aspect of the modernization process, and therefore would lose at least its public role.

As imagined by American scholars, the process of modernization would amount to little more than the all-embracing Westernization of the entire world. Although the time lags would be difficult to calculate, in the end we would witness an irresistible convergence of all societies and states upon the model of the most advanced industrial nations of Europe and especially of North America. The driving force behind the process would be economic modernization, understood as industrial capitalism, rendered inevitable due to the pressures exerted by global competition on every country. Economic modernization, in turn, would bring about an across-the-board Westernization of the rest of society and its principal institutions. If one asked for some palpable evidence that this process was really happening, nothing could be more convincing than the global triumph of American pop culture and its most famous icons such as Walt Disney, McDonalds, Coca Cola, and Hollywood, which took the world by storm from the 1950s onward, except in the places where their spread could be halted by violence.

The first major misgivings about the explanatory power and normative persuasiveness of this model arose in political science circles, but were already apparent as facts on the ground as early as the 1950s. In the wake of the broad wave of decolonization in Africa and Asia, it suddenly became apparent that political institutions imported from the "civilizational sphere" of the West's erstwhile imperial powers would have quite different outcomes in their former colonies than what had been intended. In some cases, they simply did not work at all. Either way, the reason for their failure seems to have been that these countries had entirely different cultures than those in which the imported institutions had originated. In some of the countries of Asia and Africa they became empty shells that subsequently were filled in by quite different practices drawn from the traditional way of life. In other cases the imported institutions simply were rejected altogether. The traditional culture in these countries proved to be much more resilient and influential than the modern institutions imported by the colonial powers that had dominated, in some cases for centuries, even though the latter had tried hard to influence local elites in favor of their Western ways.

The unexpected yet obvious mismatch between local and imported institutions turned out to be tenacious and durable. As a response to the challenge it presented, an independent branch of political science research began to emerge in the 1950s. The aim of this literature was to understand the origins and distinctive characteristics of the respective regional and national traditions and the political cultures allied to them and to see how they reacted to, and in turn reshaped, institutional transformations. In a certain sense it was an early form of empirical research into multiple modernities, epecially since it readily became apparent that the political culture of a given country could not be anything but the application to political arrangements of the country's general historic culture. The pragmatic research authored by US political scientists Gabriel Almond and Sidney Verba, *The Civic Culture* (1963), pioneered this approach and remained an influential frame of reference for such investigations for many years. At any rate, researchers who study political culture since have reached a consensus on at least one key point. Political culture, as one aspect of specific tradition-bound great cultures, as a rule will be quite successful in resisting rapid changes demanded by the importation of "Western-modern" institutions. In some instances – which ones is a difficult matter to predict – they will be able permanently to defy the expected transformation. Consequently we are now better able to understand how a given society may either adapt imported institutions, often adjusting their actual use to the structures that have emerged from their own cultural traditions, or else reject them entirely. The utility of the study of political cultures has been confirmed by numerous wide-ranging empirical investigations covering almost every regional culture across the globe, Yet for an amazingly long time it had scarcely any impact upon classical modernization theory.

The heyday of research on political culture, which in its own fashion made great strides toward recognizing the reality of multiple modernities, has passed by now, although the same cannot be said of its results, which remain highly relevant. Nevertheless, many scholars in the cultural and social sciences are increasingly inclined to think that, in some ways, the classical convergence approach (universal modernization in accordance with the Western model) has been outstripped or flat-out refuted by actual events in important areas of the world. The trouble with that approach is that on crucial questions, it essentially limits the possibilities for understanding cultures, societies, and political systems in a globalized world. The pivot toward a different understanding of modernization was first given powerful conceptual form by the Israeli sociologist Shmuel Eisenstadt. His term for it, *multiple modernities*, has had great resonance among scholars. He argued that the notion and the core content of modernity as understood in the tradition of Max Weber is open to a broad variety of differing and even competing interpretations and embodiments, all of which have or should have equal value and legitimacy. Thus, Eisenstadt's approach claimed to replace the hitherto dominant theories of modernization, i.e., those that contend that cultural differences do not matter substantially as shaping forces in the universal process of modernization in which sooner or later all countries are going to embrace the Western model. Meanwhile, by consensus among scholars, the "multiple modernities" approach to political development can lay claim to the terrain that extends between two controversial paradigms in this field of study: one represented by *The End of History* (Fukuyama 1992) and the other by *The Clash of Civilizations* (Huntington 1996). The former asserts that the collapse of communism eliminated the only credible rival ideology to Western liberal democracy. If that were the case, then there would be only one modernity, that of North America, Western Europe, Japan, and certain British settler societies, and it would be destined to spread around the globe as other countries, by adopting its underlying methods, values, and institutions, gradually "catch up" to the West. By contrast, Huntington's paradigm advocates a profoundly different take on the world's future. He identifies ten or eleven essentially distinct and often hostile civilizations, each featuring a more or less shared culture, including ethics, religion, and sometimes ethnicity. These civilizations are likely to clash in the future, especially along the "fault lines" that separate their member states (e.g., the India–Pakistan border, orthodox Russia and its Muslim and Western Christian neighbors). Western institutions and values will be rejected by other civilizations as Trojan horses opening the way for Western domination. Their universalistic aspirations and claims will be denied. Huntington's approach freezes the civilizations around an essentialist core, and rules out the possibility that they might evolve, adapt, and find common ground.

The Fukuyama–Huntington controversy has found echoes in wide-ranging philosophical debates over "multiculturalism" and the degree to which other cultures (whether within Western societies or in other countries) should be treated as monolithic and thus "essentialized," or whether their values should be interrogated and, if necessary, overridden by presumably universalistic standards such as human rights. Amartya Sen, for example, argues against the view that "Asian values" are inherently authoritarian. Citing a variety of Chinese and Indian sources, Sen points out that allegedly Western values such as toleration and dissent have long lineages in Asian thought as well (Sen 2000b: 231–38). It therefore would be a mistake to let Asia "off the hook" when it comes to enacting and enforcing human rights. Human rights have potentially universal validity; they are not merely Western inventions. Consequently, all countries should respect them. Similarly, Seyla Benhabib criticizes the "reification" of group identity and "freezing of group differences," urging instead that a "deliberative democratic model" would permit "maximum cultural contestation" within the public sphere (Benhabib 2002: ix). Both authors anticipate and enrich the multiple modernities perspective by looking for ways to respect other cultures without treating them as immutable and beyond critique.

The multiple modernities approach explores the terrain between these two alternatives. It recognizes that almost every society, regardless of which civilization it belongs to, wishes to be modern – at least in some respects. But because the desire to modernize finds expression within quite different cultural contexts, it may turn out that modern societies will never converge on a single model. Instead, they will remain culturally distinct as far ahead as the eye can see. As Eisenstadt remarks in his by-now classic essay, "One of the most important implications of the term 'multiple modernities' is that modernity and Westernization are not identical" (Eisenstadt 2000: 2).

What makes Eisenstadt's still-disputed approach promising is that it does not stipulate anything close to Huntington's relativism. Instead, it presupposes a common core of all the different types of modernity. Eisenstadt's proposal for the definition of such a core has been in the focus of recent academic debates concerning his theory. He sees the key difference between modernism and traditionalism in "the conception of human agency, and of its place in the flow of time." Modernism embraces an idea of the future that is characterized by a number of alternatives realizable through autonomous human agency, or the principle of subjectivism. The premises on which the social, ontological, and political order is based, and their legitimation, are no longer taken for granted as objective or given. Therefore, "modernity and Westernization are not identical; Western patterns of modernity are not the only authentic modernities, though they enjoy historical precedence and continue to be a basic reference point for others" (Eisenstadt 2000: 2–3).

In the framework of the U.N. and in academic discourse, four new analytical concepts have emerged, each of which – much like the literature on multiple modernities – claims to represent a non-Western political perspective on modernization. These approaches to development go by the names "good society," "good governance," "human security," and "varieties of capitalism." The main objective of the present anthology is to determine whether and to what degree such concepts enable us to discover commonalities and differences that distinguish the emerging multiple modernities in our time, particularly with regard to their political implications and consequences. Do they advance shared visions that are pursued or have the potential of being pursued across cultures and to assess the progress of modernization without a Western bias? Do they form a platform that will enable us better to understand the prospects for unity and to offer a new forum for discussions among different societies? The essays contained in this volume discuss these topics from the vantage point of political science. Within that disciplinary structure, the present volume represents an innovative turn in the discussion, advancing new arguments and raising new questions for further research. It is both of theoretical academic and of practical political interest for understanding the most basic developments of our time.

There are two competing approaches to researching multiple modernities: the conceptual and the empirical. Both are well-represented in the two previous volumes devoted to the topic (Sachsenmaier & Riedel 2001, Eisenstadt 2002). The first seeks to ascertain what modernity is and to pin down the ways in which modern societies differ from pre-modern ones – i.e., to determine what constitutes the "common core" of modernity. It builds on a long history of attempts by social scientists to define modernity in terms of characteristic institutions, beginning with Max Weber's concepts of rational-legal authority and bureaucracy (1968). By contrast, the empirical approach studies particular societies or even civilizations in search of un-Western ways of assimilating and refashioning modernity. Examples would include Islamic countries, India, or China. All of these cases suggest that the process of modernization does not play out in a vacuum. As modern practices are assimilated, they are altered and given a local or national flavor in the countries that have adopted them, so that, for example, democracy works rather differently in Japan or India than in Great Britain or the United States. Empirical reflections on the uniqueness of each country caught up in the race to modernize show that modernity is indeed multiple in the sense that it is never simply "copied" from one country to another. There is never complete convergence on one model of modernity. Moreover, modernity mainly occurs in a global arena in this day and age, one in which all societies mutually influence one another, while still reshaping or preserving many of their unique traditional features to varying degrees.

Both supporters and critics of the entire multiple modernities approach have raised the question of how exactly to conceive projects that legitimately can claim to be modern. A consensus developed to define modernity in an abstract, non-institutionally-specific way so as to free it of the neo-colonial, time-bound taints that prevailed in the 1950s and 1960s. Eisenstadt, as mentioned above, distilled his "common core of modernity" from earlier figures such as Weber, and associated it with "autonomous human agency" or the project of reshaping society through "reflexivity." Other scholars tended either to characterize this approach as a series of "promissory notes" (Wittrock 2002: 36) or to propose a more precise and substantial definition, such as "the idea of future-oriented progress and individual emancipation" (Göle 2002: 92). In partial agreement with these scholars, Anthony Giddens contends that "reflexivity" is the essential characteristic of modernity, and that "modernity is itself deeply and intrinsically sociological," in so far as sociology is a kind of continuing reflection about or profile of society and its possible future (Giddens 1990: 38–43).

But there is a price to be paid for defining modernity in such a "soft" way: it allows nearly everything to qualify as modern. For example, the Soviet Union during its formative period under Stalin was about as far removed as you could be from the democratic welfare state depicted as the model of modernity by thinkers like Edward Shils (as detailed below). Yet Stalinist Russia did fulfil, albeit through terror and repression, some of the criteria set forth by many theorists of multiple modernities. The Communist Party set out to remake Russian society and economy from the ground up, on the assumption that human autonomy would prevail over inert tradition and that a "new Soviet man" could be crafted from the clay of Russia's past. At least one scholar, Johann Arnason, has drawn the conclusion that the "communist experience" should be understood as an "offshoot of the global modernizing process": in other words, that it "can be located on the spectrum of multiple modernities" (Arnason 2002: 61). In short, the problem with highly abstract and plastic concepts such as the ones we have just considered (modernity is all about autonomy and reflexivity) is that they lend themselves to conclusions that, to say the least, do not seem intuitive.

Other scholars have inquired whether the multiple modernities approach might be the symptom of a larger shift in sociological theory. The advocates of classical modernization theory in the 1950s and 1960s assumed that modernization had its own inherent logic, sedimented in the principles and practices of individual autonomy and mastery and expressed in the differentiation of structures and the rationalization of all aspects of life. If a society failed to get on board the modernization express, it would be left in the dust by those that did. But by the 1980s, due to factors such as the economic crises, protest movements, individualization (the breakdown of encompassing, integrative institutions), and globalization, many intellectuals held that modernity itself would be superseded by a postmodern condition, one that might defy sociological understanding. In that case, the multiple modernities approach would be an attempt to revive sociological theorizing by admitting that the latter cannot provide a single model of what a modern society should look like, and thus would have to study empirical conditions, especially cultural legacies, in diverse societies (Wagner 2001).

The decisive point here is that the analytic content of modernity's common core turns on the meaning of multiple modernities. What do such general criteria as subjectivity, reflexivity, and difference imply for the political process and its essential outcomes? Can we arrive at sufficiently clear distinctions between modern and non- or anti-modern phenomena? What crucial changes occur in the political process of traditional societies that signal that they have started down the path of modernization? Even social scientists from non-Western countries readily admit that modernization causes wrenching changes in both the organization of society and the mental world that the locals formerly had inhabited: e.g., in challenging previously unquestioned religious beliefs and forcing their advocates – perhaps for the first time – to defend them on rational grounds against critics who do not share them. Couldn't one argue that modernization might eventually bring about the long-awaited convergence, but after a much longer transitional period than might have been expected?

In a 1959 address, Edward Shils summed up classical modernization theory by asserting that, to be modern, societies had to be democratic, egalitarian, scientific, economically advanced, and sovereign, all of which would require them to be "welfare states" much like the post-New Deal United States. He added that to be "modern means being western without the onus of following the West" (quoted in Gilman 2003: 1–2). Many interlocutors in this debate have followed his lead, publishing studies concerning Islam (Göle 2002: 91–117, 119–35), India (Kaviraj 2002: 137–61), or China (Eisenstadt 2002: 195–218) and modernity. Other scholars have contributed to this project though empirical work on modernizing countries (King 2002: 139–52, Wakeman 2001: 153–69).

After all, while advanced societies such as Japan and South Korea are by no means Western in every respect, most of their major institutions – parliaments, elections, courts, the bureaucracy, universities, research institutes, hospitals, transportation networks – at least superficially resemble those in Western countries. Of course, this generalization requires some qualifications, especially concerning Japan; there, the party system and family-oriented political culture that dominates the recruitment of personnel both support a form of government in which, for many decades, a de facto one-party system prevailed, and in which the governing party was able to rule in a quasi-authoritarian manner, even while doing so in an apparently Western framework.

Classic modernization theorists in the mold of Edward Shils, Walt Rostow (1960), or for that matter Francis Fukuyama (1992) would say that modernity, like the science and technology that form such a crucial part of it, requires adopting rational forms of organization and continuing to refine and improve them. Dysfunctional practices and beliefs eventually will get scrapped. They might even charge that champions of multiple modernities sometimes seem inclined to make excuses for countries that continue imposing authoritarian, repressive policies on their citizens, justifying those policies by calling them traditional (Sen 2000).

Perhaps the best answer to critics of the empirical version of multiple modernities can be found in a revised functionalist theory. One could argue that the central feature of a modern society is its ability to keep open channels of communication so that bad policies can be recognized and eventually changed. Certainly, one function of Western-style democracy is to let ruling elites know how ordinary people are faring under government policies, whether through elections, opinion polls, demonstrations, town meetings, civil society, a free press, or other institutions. Governments that block all bottom-to-top channels of communication and participation hardly can be said to meet the standards of modern political development, and may be blocking off the pathways toward change and further modernization. But are there different ways of achieving the same ends? Can a society pursue the "modern" goal of unimpeded communication and active popular participation without adopting all the Western institutions that usually accompany it, by devising functionally equivalent practices? The case of Singapore suggests that it can (Mauzy & Milne 2002). As is well known, Lee Kuan Yew, Singapore's Prime Minister for nearly 30 years, rejected equating modernization with Westernization and sought a road for Singapore that would be based on "Asian values." In practice, his model meant that open criticism of the government, especially by a hostile press, was frowned upon, and that elections of questionable fairness usually produced lopsided majorities for his People's Action Party. On the other hand, the party has tried to open channels of communication and participation at a more local level, for example though the unique Feedback Unit of the Ministry for Community Development. The Unit first decides which individuals should be considered "opinion-makers" and assigns them to some 27 feedback groups. They are then invited to attend dialogue sessions on particular issues. As then-premier Goh Chok Tong remarked, the government was interested in consulting "with people who have good knowledge of the subject," as opposed to the man on the street. As Milne and Mauzy conclude, the point of the Feedback Unit is to get information and intelligence from the people, not necessarily to promote democracy as such. This would certainly be an example of a Lee Kuan Yew-style application of "Asian values" (here extraordinary respect for learning and expertise) to problems of governance.

Similarly, in 1997 the government of Singapore (i.e., essentially the P.A.P.) adopted a plan called Singapore 21, which was designed to encourage the emergence of a more active civil society and greater participation, not because it thought that democracy was inherently a good idea, but to prevent discontented Singaporeans from emigrating (the dreaded "brain drain"). Its watchword was "active citizenship." More generally, Lee Hsien Loong, the country's third PM, has tried to clarify the Singaporean approach to public communication and participation, stating that although citizens never should fear to question a policy, they must not attack the government or disparage its fitness to rule. Singapore is effectively a one-party state and is ranked as only "partly free" by Freedom House, mainly because of tight informal restrictions on the press, and therefore it would not count as modern by the standards of Shils and other modernization theorists. Yet it is very successful in most measures of "output legitimation," having achieved a high standard of living, an apparently quite stable political system, and a supremely orderly society. It would be frankly ridiculous to say that Singapore is not truly modern. The country has met many of the functional imperatives of modern politics and social organization in ways not commonly found in the West. For example, if we look for functional equivalents of Western-style participation and communication, we will find them – at least some of them. In short, Singapore suggests a way in which even semi-authoritarian regimes can still be modern. And there is no indication that Singapore is "converging" on American- or British-style democracy. The example shows that non-Western political practices and institutions need not be fig-leaves for old-fashioned despotism; they can perform some of the same functions that parallel Western institutions do, albeit in quite different ways.

The present volume incorporates the latest research while continuing the debate on hitherto neglected ground: politics, in which the core issues of the multiple modernities approach can be revisited in perhaps their most important dimension. Other research areas in a field that has seen a proliferation of studies are not included in this volume, even though they have put the analytic value of this approach to the test across a broad array of disciplines, such as theater studies, religious studies, fine arts, and intercultural communication. Still others build on the multiple modernities approach to reconsider Karl Jaspers' seminal work on the emergence and content of the axial age, the time around 500 BCE, when some of the world's great religions entered history and philosophy arose in ancient Greece.

Sociological studies on the subject deal with a variety of fascinating questions. These include the transformation of the political agency of religious groups within transnational civil society under the globalized conditions that have weakened the sovereign nation-state; how local, national, transnational, global, and virtual spaces ought to engage in intercivilizational discourse without presuming secular assumptions tied to cosmopolitanism; and the limits and modes of trans-cultural communication. Regional case studies covering Africa, the Middle East, Russia, and South America likewise examine the role of religion in the process of multiple modernization. Such research evinces great interest in the question of whether modernity, democracy, and secularism are universalistic concepts or are unique to Western civilization (Stoeckle & Rossati 2018). Some recent studies in the field analyze how the concepts of modernity and postmodernity are perceived in different cultures. Several of the latest books on the topic specialize in issues of art and its transcultural performance and perception (Preyer & Sussman 2015).

Yet none of these more recent publications takes a distinctively political-scientific approach. The volume presented here exclusively contains contributions by international political scientists and is dedicated to questions of governance, government, and international relations. In this respect it differs markedly from competing books on the general topic of multiple modernities (Bowman 2015).

Although the current state of research displays a high level of development and internal diversification, the debate over the meaning and worth of the multiple modernities approach continues. Authors whose viewpoint continues to be grounded in the theory of a uniform, Western modernity have responded to the debate within the multiple modernities field by attempting to revitalize and refine their own ideas. A good example of that tendency is found in the contribution of Jack Snyder to this volume. As yet, political scientists generally have not engaged in sophisticated discourse about the points of agreement and disagreement between the two paradigms.

Because of this gap, one main focus of this book will be on the consequences that the multiple modernities approach might have for comparative research on political systems and political cultures, especially in the realm of democratic theory. The book's structure is aimed at promoting an intercultural scholarly dialogue on a few selected thematic areas, not so much as a conversation between authors as a series of common references to the same topics.

That characterization fits the contributions on good governance in Part I. *Thomas Meyer* argues that a more precise concept of good governance provides an indispensable supplement to multiple modernities as a means of distinguishing among the various paths toward political modernity and preventing a slide into bottomless relativism. The question posed by *Rodney Bruce Hall* ("Whose Modernity?") and the answers he provides to it embody a principled skepticism about whether it is possible to find a reliable criterion to evaluate contradictory claims concerning modernization.

In Part II, *Jack Snyder* presents one version of new modernization theory that takes into account the critique elaborated by the multiple modernities approach. By contrast, *Tak-Wing Ngo* points to China as a unique variety of capitalism clearly distinct from the varieties that have been shaped by the Western tradition. In its essential features, the kind of capitalism practiced in China is closely tied to Chinese cultural traditions and shows no signs of being a transitional form that might converge with the "Western" varieties of capitalism in the foreseeable future.

In Part III, *Amitav Acharya* and *Inge Kaul* rely on different perspectives to explain that the concept of human security in both its origins and meaning is well-suited to provide a demonstrably non-Western yet universalistic criterion for assessing progress toward modernization and the success of political action.

Part IV focuses on the function of dialogue in relating multiple modernities to each other. *Mario Telò* reconstructs the historical paths that took intra-European culture from conflict to dialogue with its inclusive effects, while *UN High Representative Nassir Abdulazis Al-Nasser* reports on the goals, problems, and experiences of the United Nations' project, "Alliance of Civilizations in Building Culturally Inclusive Societies in the 21st Century." *Julia Tao* explains how the Confucian doctrine of harmony might promote an inter-cultural dialogue on issues of governance and human rights.

The topic of Part V is globalization. *Manuel Castells* regards globalization, especially the trans-border financial networks it has fostered, as the defining feature of modernization. Yet he is aware that those networks have spawned anti-globalization networks of their own that resist cultural homogenization and ensure that local cultures will not converge on one single Western model anytime soon. *Julian Nida-Rümelin* reflects on refugees as migrants, a problem that has become crucial to intercultural relations and modes of existence. In effect, migrants have become ambassadors between cultures. Finally, *Jürgen Kocka* develops a trend-setting proposal for the process in which a possible basic consensus (or "core") might be negotiated among the global varieties of modernity.

PART I
Good governance

1

MULTIPLE MODERNITIES AND GOOD GOVERNANCE

Thomas Meyer

Modernization theory refuted

Since the end of the "clash of ideologies" era in the 1990s, the overall situation in the world – culturally as well as politically – has become increasingly fragmented, confused, and conflict-laden. This disarray includes a variety of phenomena: more armed struggles and all sorts of old and new wars between groups, the exact nature of which is often difficult to make out. Misunderstanding, disorientation, and suspicion nourish prejudice, fear, and sometimes aggression, all of which make the world more unsafe. In many cases, these circumstances militate against understanding and cooperation, the most urgent necessities in a time of globalization.

Before this "new obscurity" phase (Habermas 1985) of the world situation began, the global pattern seemed rather well structured. Identity and difference, belonging and opposition (both culturally and politically) seemed unambiguous: the "free West" here and the authoritarian communist camp over there, with the neutral countries more or less affiliated with one or the other of the ideological camps. Finally, an all-embracing process of modernization seemed to be underway that would guarantee a final convergence of the different systems due to either the immanent logic of technological and economic progress (Rostow 1960) or the obvious cultural superiority of the West (Lerner 1958). Its own institutional setting appeared to the West as the obvious normative model for the rest of world.

The democratic revolution in Eastern Europe in 1989 has been interpreted in two different ways. First, we have Samuel Huntington's theory, predicting that the world should prepare for a new age of insoluble conflicts since the great ideologies of the 20th century finally have lost their credibility (Huntington 1996). From his perspective, understanding, mutual recognition, and trust-building among the great world civilizations seemed unattainable because of the deep fault lines that cleave apart their irreconcilable basic values. An era of hostility and conflict was to be expected. Meanwhile, Francis Fukuyama's declaration of the end of history drew heavily from the classic modernization theories of the 1950s (Fukuyama 1992). He crowned the one and only true victor after thousands of years of struggle and trial and error in human history: the West, with its political system of free elections and economic system of free markets. In that view, liberal democracy and capitalism are presented as the solution to the puzzle of history for all of humanity. Henceforth, so the hypothesis goes, the process of Western modernization could complete its ineluctable historic task: disseminating its unique blessings over the rest of the world by instituting free elections and free markets in every country, East and West, North and South.

As a consequence of such thinking, Western-style modernization was carried out unhesitatingly in the wake of the democratic revolutions that took place across Eastern Europe. In most countries, the process occurred through acts of abrupt institutional change, with the expectation that comprehensive cultural transformations eventually would align these changes with the values and attitudes of the local populace. But these policies brought with them some unanticipated consequences, including the misery of large parts of the population, the temporary return to power of communist parties in some of the affected countries after a few years, and today a strong wave of new authoritarianism in tandem with novel brands of semi-fundamentalist identity politics (Meyer 2001).

In spite of all this, the theory of uniform Western modernization of the 1950s – one that had lost all traction in the interim because it did not fit with the new reality – was refurbished and reformulated as "neo-modernization theory" in the 1990s. Nevertheless, the theocracies of the World (Saudi Arabia, Iran, etc.), as well as China and Russia (the latter after its painful neo-liberal intermezzo), continued to pursue differing strategies that tried to combine (sometimes very) selective modernization with an insistence on the right to their own traditional values and practices.

Two new approaches: Multiple modernities and good governance

Thus, we are now in a situation that invites us to reconsider the combined concepts of modernization and Western-style government along with their implicit claims to be universal models for the rest of the world. During the previous two decades, two alternative conceptions have been put forward in response to this dilemma: the notions of multiple modernities and good governance. Both were developed in an effort to transcend the hegemonic Western discourse about what modernization means and what it implies politically that persisted until the end of the Cold War. Moreover, both concepts originated in an effort to understand and respect cultural differences while building a bridge of common values and objectives between civilizations (Meyer 2007). Thus, although they were developed and presented independently of each other, the intention behind them is similar and they largely overlap.

The *multiple modernities* approach initially was conceived by the renowned Israeli sociologist S. N. Eisenstadt (2000), who wanted to challenge the hitherto-undisputed monopoly of Western-style modernity as the universal form of modern culture.

Eisenstadt argued that the core content of modernity as understood in the tradition of Max Weber is open to a broad variety of differing or even competing interpretations and embodiments, all of which should have equal value and legitimacy. Once accepted as such, these diverse visions could have consequences across all realms of social and political life.

The concept of *good governance* was first developed on the basis of the experience of the World Bank (1992). It was a first step in overcoming the longstanding propensity of almost all the principal political actors and scholars in the West to consider the Westminster model of parliamentary government as the purest form of democracy, one that could and should be implemented in all corners of the world regardless of context, culture, and history. Good governance was based on the idea that the objectives of human welfare and well-being can be achieved or approximated in countries that reject the classical institutions of Western style democracy. The World Bank definition of good governance makes no reference to institutions or regimes; instead, it highlights several procedural and outcome dimensions of politics, including good government and public-sector management, transparency, and a legal framework for development. Other definitions along the same lines have opted for a more dynamic approach. For example, they might describe good governance as the process by which the interests of the entire society are served in fair ways with increasing opportunities for participation of all and the growing implementation of the rule of law (Meyer 2013).

Both concepts are designed to overcome a long tradition of West-centrism and open up some of the core values of modern culture to a broader variety of possible and legitimate, yet competing, interpretations. They transcend the alternatives offered by Huntington and Fukuyama, respectively. And, of course, they reject the total negation of all modern cultural values by fundamentalists from all civilizations, who incidentally feel very well understood and encouraged by Huntington.

A little later another "third way" paradigm gained prominence in comparative social science and the study of social systems: the varieties of capitalism approach (Hall & Soskice 2001a). Research based on this approach demonstrated that there is no such thing as capitalism or a free market economy per se, but rather a broad range of combinations among elements: markets, regulations, state intervention, social actors, and restrictions of the rights of private property. Which of the varieties prevails in a country depends mainly on its cultural context and historic experiences. For example, coordinated market economies and liberal market economies differ in degree but not in kind, and to some extent embody cultural differences. According to research produced by this school of thought, there is no prospect of convergence among the different varieties of capitalism in the foreseeable future.

Eisenstadt's generative idea

Of course, the concept of multiple modernities cannot and does not stipulate anything approaching the uncompromising cultural relativism that characterizes Huntington's view. Instead, it presupposes a common core of all the different types of modernity. Eisenstadt's proposal for the definition of such a core has been widely accepted in the debates concerning his theory. The key difference between modernism and traditionalism (here he follows Max Weber) is "the conception of human agency and of its place in the flow of time" (Eisenstadt 2000: 3). Modernism embraces an idea of the future as characterized by a number of alternatives realizable through autonomous human agency – the principle of "subjectivism," as some have put it. The premises on which the social, ontological, and political order is based, and the legitimation of those premises, are no longer taken for granted as "objective" or given. A new "reflexivity developed around the basic ontological premises of the structures of social and political authority." Therefore, "modernity and Westernization are not identical; Western patterns of modernity are not the only authentic modernities, though they enjoy historical precedence and continue to be a basic reference point for others" (Eisenstadt 2000: 3). But the core of modernity includes "the autonomous participation of members of society in the constitution of the social and political order, or the autonomous access of all members of the society to these orders and to their center" (Eisenstadt 2000: 5).

Regarding the governance process, the multiple modernities approach would still entail, among other things, the acceptance and institutionalization (in some form) of social and political pluralism and participation both at the level of debate about the common good and at the level of political decision-making. However, the core demands of modern politics might assume a variety of institutional forms. They could vary along several axes: the prevailing historical situation, the requirement that basic order be maintained, and respect for divergences in cultures and traditions.

This pluralistic notion of the requirements of modern culture overlaps in its meaning, basic thrust, and tolerance for multiple institutional settings with the essential concerns of the concept of good governance.

Varieties of modernity

Each of the three conceptual innovations mentioned above (modernities, good governance, and varieties of capitalism) made its appearance in the academic discourse of the late 20th century and has stimulated inspiring, fruitful research and debate ever since. This is certainly due to the obvious fact that the pattern of ideologies, civilizations, and political systems observable in today's world does not display any tendency towards uniform standardization anytime soon. To oversimplify a bit, we can assign countries to distinct groups representing different spheres of cultural tradition:

- "The West" (internally split into the different cultures of libertarian and social democracies).
- Religious-political fundamentalism in tandem with theocracies (e.g., in Iran and Saudi Arabia), that deny modern core values altogether.
- Authoritarian systems with representative political institutions and elections and a sphere of economic liberty (e.g., China and Russia) that claim to pursue alternative paths of modernization.

And, in addition, there is the increasing number of failing states and/or those experiencing civil war conditions featuring regional warlord regimes (mainly in Africa) that do not offer any credible legitimizing ideas at all.

The dispute about the essence and the limits of the concept of modernization focuses on the opposing claims of representatives of the "West" and of the authoritarian regimes. Two questions are of particular interest in this context. First, where exactly should we draw the demarcation line between different modes of modernization and outright anti-modernism? And second, do we have clear criteria for distinguishing countries with more or less good governance from those that definitely display bad governance? These are questions that are interesting for theoretical reasons and political assessment as well as for the choice of adequate approaches to international/intercultural dialogue and cooperation.

The logic and dynamics of modernization

Where can we look for answers? From the multiple modernities approach, we can learn that there is a common core of modern culture, even though there may be a broad variety of institutional expressions of this core. As mentioned above, the common core can be formulated succinctly as the principle of subjectivity or self-determination. The renowned sociologist Richard Münch (2001) has proposed that this core can be understood as the universal dimension, or the culturally neutral logic of modernization, consisting in the very general principles of rationalism, secularism, individualism, and universalism (in the sense that arguments address a global community of interlocutors).

It seems obvious that the process of implementing the same logic of modernization in culturally variable settings (the dynamics of modernization) will be conducive to different results depending on its tempo, the culture in question, the historical and social conditions of each country in which the process of modernization takes place, and the point in time at which we observe it. Thus, Japanese, Chinese, American, Russian, Iranian, South African, French, or British modernization all will be and will remain different. Yet, it is also obvious that none of the differences among them can be justified simply on the grounds that each is different. It must be shown persuasively that these aspects of difference still maintain some connection to the common core. In that sense, diversification is in itself the unavoidable logic of modernization, at least as long as "modern" institutions are not imposed on a country from outside. In order to ascertain whether such a connection really exists, it makes sense, at any given point in time, to ask: Is the level of modernization that has been attained so far in a given country permanent, at least according to the judgment of the relevant local authorities, and mainly due to the cultural conditions in that country? Or is it temporary (again according to relevant authorities) and mainly due to a lack of time and/or resources? The answer to this question matters tremendously, and opens up issues for further research.

To forestall serious confusion, we need to make it clear at this point that the model of the modernization process adumbrated in these pages by no means entails an essentialist notion of culture, of the kind that inevitably informs theoretical schemes such as Huntington's. The idea of a clash of civilizations and its justification by reference to static fault lines running between the basic values of the contending civilizations presupposes that cultural contexts can be treated as though they were natural phenomena incapable of any real change. By contrast, the Münch model – taken as definitive in this context – is both dynamic and open. Yet we must go farther still, treating culture under conditions of globalization (really even before that) as constitutive: i.e., as a dynamic space of social discourse shaped by contradictions, in which competing actors contend to interpret and pass on tradition in light of their accumulated experiences and existing social conditions. What decides the version of tradition that will be handed down is not the content of that tradition itself, but instead the balance of forces among the competing elites charged with interpreting it. The resulting body of tradition that such an elite succeeds in getting accepted is thus always provisional. It is constantly being questioned by major or minor actors who present more or less drastic alternatives and, in various ways, try to win acceptance for their own versions. Even when it seems as though an entire society is clinging rigidly and dogmatically to certain cultural patrimonies, closer examination shows that its elites are involved in a constant defensive struggle against ever-present alternatives offered by those with an interest in challenging prevailing schemes.

Unquestionably, when the defenders of tradition in a given cultural sphere come out on top – as is often the case – their success bespeaks a high degree of concordance between their version of tradition and the lived experience of the society, for otherwise the tensions between the latter and the claims of traditionalists would become untenable over the long run. In addition, both globalized channels of communication and changing lifeways in the society in question inevitably will cause the traditions to mutate, even though this process may happen at an almost subliminal level. Here it is worth recalling the findings of political culture research between the 1970s and 1990s. They indicate that, when a significant gap arises between political culture and political institutions, the usual outcome is either institutional change that re-establishes a rough equilibrium or else a chronic crisis (Almond & Verba 1963). Even though state or other (religious or societal) actors possessing extensive power resources can exert a degree of control over such phenomena (cultural change or cultural stagnation), they cannot create or prevent them. It follows that, as a rule, elites will have some prospect of success in invoking traditional versions of their cultural legacy to legitimize their rule only when they can count on broad social resonance. Obviously, this endeavor will turn out best if and when their procedures and outcomes dovetail nicely with traditionalistic patterns of legitimation, especially when everyone can see that they have durably satisfied the most crucial and even some of the higher-order needs of the governed, as compared to the past and to the achievements of other countries.

The antipode of the essentialist misunderstanding of culture is the nihilist error, one that is perhaps even more problematic in its social and political repercussions. We encounter it whenever people assume that cultures are reducible to the cognitive outcomes of current political discourses. Thus – so the assumption runs – however those public debates over a culture's meaning turn out, its broad outlines are susceptible to being changed almost overnight. When one thinks of culture in this way, one loses sight of the deeper social and emotional dimensions that give it staying power. Those dimensions unceasingly are reproduced and reinforced in processes of individual socialization, the entire system of societal institutions, and the rituals and practices of everyday life. The result of the nihilist confusion is a kind of constructivism incapable of recognizing that social constructions always work with traditional materials. They cannot be invented ex nihilo.

Sequencing in the process of modernization

In this respect, the question of sequencing modernization, identifying its different dimensions, and specifying their relevance for each other becomes especially interesting. Mick Moore (1993) from the British Institute for Development Studies has delivered a profound critique of the original World Bank conception of good governance. He proposes to transcend the typical Western list of the basic challenges of governance by looking more closely into the situation of third world countries. Building on his argument, it seems useful to distinguish among three types of key challenges that societies may face at different stages of their development that require different political answers depending on their cultural resources.

- Establishing social peace among different ethnic or other groupings by way of building a stable political order offering at least minimal legal protection for all citizens. This task relates to early phases of nation-building and often may remain a permanent challenge in heterogeneous nations.
- Abolishing poverty and producing a sufficient level of wealth that is shared reasonably widely by the entire society.
- Enabling sufficient levels of opportunity for all citizens to participate, particularly when the level of wealth and education for large sectors of the society increases.

Two fundamental questions derive from this list. First, is there an optimal sequence for meeting these three basic challenges in the reality of social life? That is, are there stages of development that must be passed through before the next stage can be attempted, or can they all be transited simultaneously? And second, what role does the time factor play?

If we examine the European process of modernization, we find that nation-building and the creation of a stable legal order for the whole nation were not achieved anywhere under democratic governments. During the 18th and 19th centuries, by which time national unification had largely been achieved throughout most of Europe, diverse forms of authoritarianism prevailed everywhere.

Then, once a functioning legal order had been maintained for an extended period of time (often one and a half centuries), democratization – i.e., increasing political participation by a growing number of citizens – was won in highly conflictual processes that lasted in some countries until quite recently.

Finally, the increasing production of wealth linked to a more or less inclusive pattern of wealth distribution and social security was not achieved in any democratic country in the West until the 20th century, often not until the second half of it.

Thus, the whole process of "modernization" in the West took roughly two centuries.

The classical sequence of Western modernization in Europe started in the second half of the 18th Century and lasted until the end of 20th century after passing through the following stages:

- Nation-building
- Instituting the rule of law
- Industrializing/creating wealth
- Democratizing
- Establishing a welfare state

It is noteworthy that democracy arrived rather late in this sequence and was introduced gradually within the framework of the rule of law. The rule of law, in turn, developed under authoritarian conditions, in a process that took more than a century. By contrast, most developing countries face all these basic challenges at the same time. What is more, they often must deal with particularly difficult conditions and often persistent problems in the dimension of nation-building. Finally, they sometimes face inflexible time limits, insofar as they are expected to modernize across all five dimensions in just a few decades.

Basic rights, multiple modernities, and good governance

One of the especially interesting and innovative elements in the concept of good governance is that it synthesizes both sides of the political process: input and output. This characteristic can also be conceived of in terms of a two-dimensional model of political inclusion comprising both the political process and its outcome. In contrast to the usual political regime approaches, the concept of good governance aims at inclusion in regard to both input (participation) and output (level and distribution of resources). If inclusion characterizes one pole of the political process while exclusion prevails at the other, then it does not deserve to bear the name of good governance. If only one of those poles is (sufficiently) inclusive, the classification has to be mixed: neither altogether good nor altogether bad governance. If both poles are predominantly exclusive, then the assessment is unambiguous: bad governance.

At this point it is advisable, if not unavoidable, to bring into play the basic rights guaranteed in the U.N. Covenant of 1966. One of the most conspicuous features of both the modernities and the good governance approaches is their lack of reference to universal basic rights. Neither Eisenstadt and his followers nor the key authors of the good governance concept (like the World Bank) mention them. Yet, there is an internal relation between good governance and basic rights because of the comprehensive nature of both. They each embrace both sides of the governance process – form and content, input and output – in a similar spirit. It is often overlooked that the Covenant on Basic Rights is not just a moral declaration of good will, but a valid part of international law. It consists of two parts, liberal basic rights and social basic rights. This makes it a convincing expression of the fundamental conditions of human agency and reflexivity and thus likewise of the core principle of modernity.

The preamble to the covenant makes it clear that the principle of self-determined action requires guarantees for both negative and positive freedom: the effective rule of law that protects the conditions of freedom as well as a guarantee of access by every citizen to the material and social resources needed for self-determined action. If freedom is to be fully realized, then neither of the two categories of rights can be omitted. Consequently, both are of equal moral value and legal status. They are considered as interdependent and mutually supportive. If a person has the formal right to free action, but lacks the material resources to make use of this right, then that person cannot be said to enjoy the real opportunity to lead a free, self-determined life.

The conclusion to be drawn from this line of argument is that a transculturally valid yardstick to assess good government asks whether the ensemble of basic rights – those concerning the political process and those concerning the quality of its social and economic outcomes for all – has been respected. It is obvious that the mere institutionalization of civil and political rights, i.e., of a Western-style political regime, does not guarantee good governance output, and neither does the validity of social and economic rights if they are not complemented by the rule of law and the opportunity to participate. Only the application of both categories of rights in tandem qualifies as good governance in the full sense.

Obviously, the achievement of just a part or a certain degree of one or both categories of rights by a country poses a challenge for how to evaluate its progress. But the main point in this trans-cultural approach is that the mere existence of a set of institutions neither constitutes not guarantees good governance in and of itself. Moreover, the absence of the classical set of Western political institutions is not tantamount to completely bad governance.

Human agency rather than institutional determinism

For many developing countries, modernization is not a transition from one system to another but an open process of trial and error, of finding out what suits the country best. Intellectuals and politicians in third world countries know that they are participants in the worldwide process of modernization and therefore find themselves in an inevitably reflexive situation. Unavoidably, they have to draw upon the resources inherent in their particular traditions from a modern/reflexive and not from a traditionalist point of view. They must relate to the state of modernization in the West in full awareness of its crises, undesirable outcomes, and dark sides. Finally, they must provide feasible solutions to the immediate social, economic, national, and political challenges facing their countries.

It is worthwhile to try to understand this reflexivity because the course of modernization is not an objectively determined event or an auto-dynamic (auto-poietic) mechanism. Moreover, it imitates the West only in rare instances. It is in most cases enacted by a set of relevant actors, a matter of human agency and therefore of choice. The actors usually know about the changing fate of modernizing societies, about unanticipated or undesired consequences of present decisions, and about social or political costs that have to be paid for them. And, above all, they know about the risky challenge of organizing legitimacy and mass support for the modernization path which they intend to choose in their own countries.

It would be informative to compare India and China in this respect. There is insufficient space to do so here, but a brief glance may be cast on China's claim to be a modernizing country that is following its own uniquely Chinese way. Ever since the middle of the 19th century (if not earlier) there has been a debate in China about how to select the modernization path best suited for the country (Gransow 2006: 152). In the second half of the 19th century, intellectuals in the Yangwu movement proposed to preserve the substance of the traditional Chinese culture while adopting only the best technological and military achievements from the West. In order to attain levels of affluence comparable to those of the West, some proposed in addition to copy key elements of its market economy and entrepreneurial style. The Meiji Restoration in Japan served as another point of reference for this idea of a third way of modernization under the formula "tiyong," the traditional Chinese doctrines would serve as the substance of the new mixture whereas the Western doctrines would function as mere practices or applications (Zhang Zhidong). The new relationship would cast the Chinese part as the main issue and the Western part as a series of side issues (the "way" and the "instruments"). The overall objective for the right blend between East and West was wealth and strength for the country. The concepts of modernization and Westernization were carefully distinguished from each other. The famous reform movement of 1898 rejected this concept but shared the overall objective of wealth and strength. So did the so-called Movement of May 4, 1919. Both argued that China would have to adopt the full program of Westernization in order to achieve the goals of wealth and strength.

In some respects, this discussion has continued ever since. Even Deng Xiao Ping's concept of the Four Modernizations (1978) reflects it. The latter was understood as a way to carry out partial or selective Westernization in the fields of agriculture, industry, science and technology, and military/defense. Yet, in each of these selected fields, what actually happened was more like a synthesis between Chinese tradition and Western achievements than pure Westernization.

The choices that the principal actors make in a given situation is not institutionally determined in the sense of classical modernization doctrine, which presumed that once you buy into one of the modern institutions, you will have to adopt the complete Western set in order to make the arrangement work. In this regard, the decision of the Chinese leaders about how to pursue the kind of modernization they prefer is perfectly modern, according to the doctrine of Eisenstadt and Weber. But is it also modern in the sense of good governance?

Different types of authoritarian states

The two approaches we have reviewed, multiple modernities and good governance, suggest that it is neither informative analytically nor justified normatively to draw one general demarcation line between countries that stick to the Western set of political institutions and the rest of the world. A truly transnational, transcultural approach must be more nuanced and complex. Western institutions can fail to deliver fully the outcome that the good governance approach demands, while non-Western institutions can bring about a remarkably close approximation of good governance in certain areas, such as the social and economic.

The indispensable criterion of modernity is human agency: the autonomy of subjects, their right to act and to participate, and the reflexivity of the social and political order. Consequently, we must take into consideration two ways in which regimes can fall short of good governance. The first of these is the absence of civil and political basic rights, which would be tantamount to a negation of liberal principles altogether, and the second is the absence of social and economic basic rights, i.e., the denial of the social supports that sustain the former. Illiberality has a strong relation to the input side of the governance process, while social and economic rights pertain to the output side (see Table 1.1). By these strict criteria, only a group of European countries, plus a few from Asia and Latin America, deserve unqualified endorsement as exemplars of good governance and – a fortiori – as fully modern. Almost all other countries fail to meet either the liberal or the social criteria or both.

Illiberal (authoritarian) rule can take very different shapes and produce widely divergent results. In some countries in which the very maintenance of public order is a pressing need, authoritarian rule may be rooted in strong and popular cultural traditions, as is the case in Russia, Japan, China, or India. If authoritarian rule in such cases is exercised by responsible elites pursuing policies of social and economic inclusion and the common good of the whole society, some of the output criteria for good governance may be met. By contrast, in many African countries authoritarian rule is only instrumental for securing certain clan or coterie interests and ignores the wider interests of the society. Public security, rule of law, infrastructure, education, health, and social security are completely neglected. This is tantamount to a total breakdown of good governance. An analytical or normative concept that is not able to illuminate and assess such tremendous differences is not really informative.

TABLE 1.1 Varieties of political regimes

Democratic regimes	
Input – Inclusive:	a) Output – Inclusive
	b) Output – Exclusive
Authoritarian regimes	
Input – Exclusive:	a) Output – Exclusive
	b) Output – Inclusive

Strategies for postponing crucial elements of good governance or realizing them only in certain sectors

Universal basic rights can serve as critical yardsticks for assessing the degree to which good governance is being practiced in a country. One of the empirical puzzles connected with the multiple modernities approach consists in the difficulty of identifying to what degree the individual particularities of different countries in their path of development are unavoidable delays in, deliberate circumventions of, or lasting and substantial deviations from such fundamental rights.

Returning to the Chinese example, it must be said that the country's compliance with the U.N.'s list of basic rights remains ambiguous and volatile. Until just a few years ago, the argument advanced by Chinese scholars was often something like this: We already subscribe to the social and economic rights of the 1966 U.N. Covenant and will support the civil and political basic rights in a few years. Three separate reasons were presented to justify the postponement of the institutionalization of the liberal basic rights: (1) that there is a need to overcome poverty and the lack of education first; (2) that in a country without any democratic tradition and experience, there must be a slow and experimental bottom-up process, beginning with cautious democratization in the villages and urban districts; and (3) that a country in which disorder, instability, and civil war were common features throughout history needs to avoid these dangers by acquiescing in a certain degree of authoritarian rule.

Recently, the grounds offered for the delay in guaranteeing basic rights have begun to change. More and more, the preferred argument is that the key justification for government must derive from traditional concepts of political rule in the country, such as "all under heaven" and "harmony," rather than from Western ideas and demands. Only this course, so the justification runs, can guarantee stability and the rapid growth of inclusive wealth. Clearly, although such a claim is open to criticism, there is an obvious strong point in its favor: the successful elevation of some 250 million people out of grave poverty in just three decades. However, the Chinese way of legitimizing their governance process has become increasingly ambiguous. Three different justifications compete: traditional governance ethics (Confucianism), the new nationalism, and an unclear (but not entirely suppressed) invocation of the U.N. covenant on basic rights. De facto, the norm that guides the practices of the ruling party is tradition: The well-being of the entire population, harmony, and the greatness of the nation.

Hence, it seems reasonable from both an analytical and a normative point of view to distinguish between two superficially similar regime types. Those that rely principally on traditional norms and values to legitimize their power and governmental actions while also delivering universally accessible, unimpeachable service outputs as judged by any conceivable standard of value (e.g., eliminating poverty, raising living standards, ensuring social welfare provision and access to education), and making serious efforts to establish the rule of law, should be assigned to a different category than authoritarian regimes that serve only elite interests. That

TABLE 1.2 Political regimes input-output performance[1]

Country	*Input-performance*[2]		*Output-performance*[3]	
	Rank	Score	Rank	Score
Norway	1	9.93	1	0.949
Sweden	3	9.39	5	0.943
India	*35*	*7.68*	*131*	*0.624*
Singapore	*84*	*5.89*	*5*	*0.925*
Qatar	*135*	*3.18*	*33*	*0.856*
China	136	3.14	90	0.738

Notes:
[1] This chart illustrates selectively both the complex relations between input and output performance, and the need for a better representation of output performance in an appropriate evaluation of political systems. In italics are the countries with a strong mismatch between input and output performance.
[2] The Economist Democracy Index, 2016 (scale of 1 to 10, with 10 as the maximum score).
[3] UN Human Development Index 2016, released March 21, 2017 (scale of 0 to 1, with 1 as the maximum score).

judgment is especially apt when solid empirical reasons exist to think that the former regime type enjoys the consent of the majority (see Table 1.2). Nevertheless, even in this case one cannot waive the strict application of the transculturally valid norm that dissidents and their right to criticize the regime publicly deserve to be treated in ways commensurate with human dignity. At any rate, there is a variety of relativism that, citing the right countries have to pursue multiple paths to modernization, would cast aside all of the basic norms of legitimate and good governance. To be sure, there is considerable leeway in defining what good governments may do or aspire to do, but such an approach clearly places itself beyond the pale of what is acceptable.

One must concede to Jürgen Kocka (2006) the point that there are universal principles that constitute a common core of modernity despite the diversity of paths leading to it, and that these must be negotiated in a trans-cultural forum, not decreed from within one of those paths. Yet no matter how differently they might be formulated, the principles of good, inclusive governance have to establish limits that must not be transgressed. Otherwise, the very idea of good governance will be robbed of content.

Summary

Both for understanding today's world and facilitating global cooperation it is wise to replace the old illusions about a uniform, Western-style modernity and the final worldwide triumph of a single set of Western institutions by the more realistic concepts of multiple modernities and good governance (in tandem with good society and human security). Our analytical concepts for comparative research and debate in a world of multiple modernities need to become broader and more open but also more precise than the still prevailing idea of Westernization as the (often hidden) ultimate yardstick for the assessment of different countries and their development. It appears desirable to make these concepts more concrete and operational in order to enhance our ability to conduct more informative comparative research and make better and fairer assessments of the achievements of different countries.

2

GOOD GOVERNANCE IN A WORLD OF MULTIPLE MODERNITIES

Whose modernity?

Rodney Bruce Hall

It seems timely to be looking at the question posed nearly 20 years ago by the Israeli sociologist Shmuel Eisenstadt of whether a "general trend toward structural differentiation" in the domestic institutions of modernizing societies along with "continual constitution and reconstitution of a multiplicity of cultural programs" by social movements that are pursuing different visions of "modernity" has generated a world in which societies live in "multiple modernities." My paper will argue that, given the "number of possibilities realizable through autonomous human agency" (Eisenstadt 2000: 3), there have always been multiple modernities promulgated by different forms of states and societies, just as Eisenstadt suggests, through "continual constitution and reconstitution of a multiplicity of cultural programs... [and]... multiple institutional and ideological patterns... carried forward by specific social actors... [and]... social movements pursuing different programs of modernity, holding very different views of what makes society modern" (Eisenstadt 2000: 2).

This has been so ever since the notion of modernity began crystallizing as a coherent artifact of the Enlightenment project in Europe, even as the notion of modernity *qua* economic development has enjoyed a lingering attraction among these social actors. While the developing world has been keen to benefit from the material and lifestyle advantages of modernity *qua* economic development and modernity *qua* state control of society, it would be a mistake, contrary to what some observers claim, to argue that the almost forced interaction of peoples through the world with one another in the postcolonial era has generated a "world revolution of Westernization" (von Laue 1987). I concur with Eisenstadt that modernization is not coextensive with Westernization as a cultural program. Material wealth might be gained by incorporating Western models of modernity *qua* economic development. Moreover, the trauma of civil and domestic conflict might be quelled by adopting Western models of modernity *qua* statist control of society (e.g., sovereignty). But it has never followed that the various social movements and collective identities that emerge with surprising levels of diversity among human societies have very frequently adopted an uncritical, emulative Westernization as their solution-set to the plethora of societal challenges and aspirations that they face. When discussing the concept and its socioeconomic efficacy as an engine of global society and of history, we are always well advised to pause, to consider, and to qualify our assertions with the question of "whose modernity" we might be discussing.

Kantian rationalism and modernity

One likely reason for the tendency of those of us in the West to equate modernity with Westernization is that our notions of modernity derive from a limited corpus of texts in which our concepts of the social conditions of modernity originally arose, including especially those that propound or refine Kantian rationalism. Without these texts, neither the French nor the Scottish Enlightenments – and thus political and economic liberalism, respectively – likely would have seen the light of day as governance frameworks, nor even as more than passing normative critiques of the *ancien regime*. I argue that much of the content of modernity as it is envisioned by North Americans, Europeans, and the English-speaking scions of Great Britain is strongly derived from Kant's moral and political philosophy. In particular I argue that we derive our notions of the categorical imperatives of civilized behavior, our notions of a universalistic, moral, cosmopolitan culture, and finally the notion of a pacific republican union, all three, quite directly from a few of Kant's essential works.

In the Western conception of modernity, the notion of the categorical imperatives of civilized behavior come down to us in more or less direct form from Kant's *Metaphysics of Morals*. The three imperatives are given as assertions regarding the imperative of universality and the law of nature, the imperative essence of humanity, and the imperative of autonomy. Accordingly, the individual (and by aggregation, the state) is to act only according to that maxim whereby s/he can at the same time will that it should become a universal law without contradiction. One is further enjoined to act in such a way that one treats humanity, whether in one's own person or in the person of any other, never merely as a means to an end, but always at the same time as an end. Finally, a third practical principle follows from the first two as the ultimate condition of our harmony with practical reason: the idea of the will of every rational being as a universally-legislating will (Kant & Reiss 1991: 143, 155, 159). The categorical imperatives form the backbone of Kant's moral philosophy and largely have been incorporated into the baseline architecture of Western notions of the preconditions for civilized modernity.

In his essay, "The Idea of a Universal History with a Cosmopolitan Purpose," Kant generates a number of propositions to buttress the claim that rationality and moral autonomy in moral, cosmopolitan modernity will conquer the self-interested individualism characteristic of what we think of as early modern Europe. Kant holds forth the promise of a yet unrealized future state that is both moral and modern. He regards the constitutional republics of his late 18th-century Europe to have attained an advanced, though still incomplete, stage of development. They are indeed civilized but not properly moral. All other societies are consigned to an inferior status because they have not even reached the state of civilization characteristic of European nation-states, let alone the status of being moral societies. Kant argued that the European nations were tending towards a federated statehood, the culture of which was to be universalist, cosmopolitan, and moral. Once they have reached that stage, they will have attained the full status of proper modernity. At a minimum, the governance of modernity must ensure peace to fulfill its moral and cosmopolitan social purposes. It is the "moral" and "cosmopolitan" social purposes of this federation of states that render it modern. Intones Kant:

> Although this political body exists for the present only in the roughest of outlines, it nonetheless seems as if a feeling is beginning to stir in all its members, each of which have an interest in maintaining the whole. And this encourages the hope that, after many revolutions, with all their transforming effects, the highest purpose of nature, a universal *cosmopolitan existence*, will at last be realized as the matrix within which all the original capacities of the human race may develop (Kant & Reiss 1991: 51; emphasis in original).

Our contemporary conceptions of globalized, liberal modernity are strongly modeled on this Kantian formulation of the nature of the social and political order that will be required for humankind to reach its fullest potential as a species. For those of us in the West and the English-speaking world, it is the source from which liberal modernity has emerged. With Kant, we recognize that this vision of modernity requires a certain form of governance, and the West has committed itself to encouraging such governance projects in both regional and international forms.

We can find that project described by Kant in his proposed peace program in his essay "Perpetual Peace." In that work Kant introduces six "provisional articles" that are of greater interest as ways of understanding what Kant felt to be threatening to peace in his day than as practical prescriptions for states that are not governed by either constitutional or dynastic monarchs. For example, we can consider the provision that "no independent states, large or small, shall come under the dominion of another state by inheritance, exchange, purchase, or donation" (Kant & Reiss 1991: 94). Kant refers to this provision as the "definitive" article that foreshadows contemporary notions of liberal modernity. Accordingly, his injunctions announce that "the civil constitution of every state should be republican" (Kant & Reiss 1991: 99ff), that "the law of nations shall be founded on a federation of free states" (Kant & Reiss 1991: 102ff) and that "the law of world citizenship shall be limited to conditions of universal hospitality" (Kant & Reiss 1991: 105ff). One must bear in mind that "republican" does not mean "democratic," but only implies "representative" governments in which the legislative and executive powers are distinct, and therefore could encompass constitutional monarchies with empowered parliaments. In the context of contemporary European affairs, it is particularly important to note that the right to universal hospitality is *not* a universal right of migration, but rather the foreigner's right to the same protection from harm that is afforded to citizens and subjects. This resulting republican union, Kant argues, would ensure peace and stimulate world commerce, whereas autocratic governments often make war for selfish purposes and even purposes inconsistent with the interests of the state's citizenry.

Whose modernity?

At its core, the Western notion of modernity is largely informed by Kant's injunctions for the behavior of a "civilized" and "moral" society that is to lead or at least participate in a cosmopolitan global order. Kant's political philosophy has been enriched in numerous ways by the addition of notions of positive rights that individuals within free societies should enjoy. Such embellishments of Kantian principles come from many sources, including the works of French Enlightenment thinkers, particularly Rousseau, Voltaire, Montesquieu, Condorcet, d'Alembert, and Diderot, as well as British contributions by Mill, Locke, and Thomas Paine ("the rights of man"), a British subject who spent much time in America. But the specific form of this cosmopolitan world order has drawn upon the reflections of the Scottish Enlightenment as well, particularly the writings of Hume, Smith, and Ricardo, which provided a rigorous critique of mercantilist doctrines and extolled the virtues of economic liberalism both within and between societies.

Regardless of any recent populist waves on the pond, modernity as a liberal world order has prevailed since at least the late 19th century, with violent intermezzos for the carnage of two world wars. Since the end of the Second World War, the English-speaking nations and Europe, among others, have held a liberal world order to be the arrangement most consistent with modernity. A liberal world order is characterized both economically and politically by liberal institutions and minimally restricted international trade, as well as by the rule of law (national law, treaty law, and international law). In this view, the globalization of economic and political liberalism has brought significant benefits, since both versions are normatively and economically superior means of delivering a wealthy, happy citizenry. Good governance consists of normative and legal commitments, institutions, and economic best practices that advance the globalization of economic and political liberalism, while state sovereignty ensures national and cultural diversity. Europe adds a layer of supranational institutions to this mix, pooling and subduing sovereignty in the interests of perpetual continental peace, a single common market, and the hope of ever stronger political union.

Russian notions of modernity, by contrast, have recently demonstrated a preference for a world order oriented toward regional or civilizational ties and affiliations, both economically and politically. In Western eyes, at least, Russia's incursions into the territory of its Ukrainian neighbor, including a flagrant military occupation and annexation of the Crimea, have demonstrated ambiguity regarding Western notions of the rule of law. The Russian position seems to favor a rule of law consistent with a "distinct Eurasian civilization." In this somewhat impoverished variant of the rule of law, the role of regional hegemon evidently dictates treaty law and trumps international law. In this view of modernity, the globalization of economic and political liberalism appears to be the entering wedge of a trend pernicious to a self-consciously Eurasian citizenry. Here the term "good governance" would embrace all those initiatives and practices that advance the strength, coherence, and construction of a Eurasian civilization dominated by Russia. State sovereignty ensures Slavic/Eurasian national and cultural orthodoxy and hegemony. This gives rise to irredentist claims against the Ukrainian Crimea and Donbass, the Caucasus, and the Baltics. The European Union and NATO appear as pernicious threats to Russia's borders and as rival claimants upon the loyalties of natural subjects of Russian regional hegemony in the Russian "near abroad." Nostalgia for organization as an "axial empire" is abundantly in evidence (Eisenstadt 2000: 7).

Contemporary China's notion of modernity clearly encompasses a liberal world order in the economic sphere. As of this writing, Xi Jinping has addressed the World Economic Forum at Davos and tried to portray China as the torch-bearer of the global liberal trading order in the face of the (yet to be demonstrated) defection of the United States under a populist presidency. But politically speaking, Chinese modernity remains oriented toward regional or civilizational ties and affiliations. To Western eyes, Chinese territorial claims to huge swaths of the Pacific Ocean as far south as the equator, Chinese domination of a resistant Tibet and less than welcoming Hong Kong, and continued Chinese insistence on a right of sovereignty and control over Taiwan do not fit well with that country's assertions that it desires to obey the rule of law. One might say that they advocate a notion of the rule of law consistent with the accommodation of other countries to the "rise of China" and the rule of a vanguard party-state standing above and consequently outside the rule of law. This refusal of the Chinese government to subject itself, and particularly to subject the Communist Party, to the law that rules the rest of Chinese society marks a discrepant departure from Western notions of modernity, and generates fundamental limitations on the quality of the relationship that China can maintain with Western democratic states, whose legal structures limit governmental power and discipline governmental agents. Yet the classical view of modernity as economic modernization clearly has been embraced by China, while the political liberalism component is firmly rejected. Tsing dynasty *hutongs* by the millions have been torn down in major cities and high-rise towers of flats full of gleaming appliances have been flung up. In this context, I am mindful of a story told me by a British colleague during my Oxford years. She led a delegation of Chinese citizens around the UK as pro bono work for the British Council and sponsored a party in her home in Oxford the evening before they left Britain. When she asked if they had any final questions before they traveled home, the spokesperson of the group replied: "Yes, why do such modern people live in such old houses?" This question is rather sobering for those of us who can find considerable charm in Britain's listed and protected historic homes. "Good governance" in the Chinese notion of modernity appears to designate those initiatives and practices that advance the strength, coherence, and continuation of the "rise of China." To the extent that a liberal global trading order counts as such a practice, it is included as part of good governance. To the extent that domestic liberal political reform challenges the rise of China – or more accurately the rise of the Chinese party-state – it is excluded. State sovereignty ensures the integrity of the party, with the state defined in a fashion that buttresses Chinese claims to Tibet, Taiwan, and the SARs. These many irredentist claims appear to be important in undergirding the government's claim to have redeemed "a century of humiliation." The Chinese government has demonstrated an ill-concealed ambition to assert regional hegemony. It might be argued that China's aggressive posturing in the South China Sea disputes can be interpreted as an expression of nostalgia for the country's resurgence as an "axial empire" (per Eisenstadt) via regional hegemony. East Asians might be forgiven for asking whether they would be expected to kowtow as they were required to do during the imperial centuries.

The Middle East and Islamic World are so diverse that it would be misleading for us to speak of them as possessing an integrated world view. For that reason, I will confine my analysis to Islamists. It could be argued that, politically speaking, they seek a world order oriented toward regional or civilizational ties with co-religionists. Economically, this order would be financed by access to energy markets run according to liberal, market-oriented rules. The Muslim Brotherhood avows quite clearly that "Allah is our objective, the Quran is our constitution, the Prophet is our leader, and jihad is our way." Islamic cultural and political units such as the *umma* and *shura* are largely incompatible with Western notions of state and society, and the Greek polis has never proven to be an adequate political forum for the governance of Islamic communities (Cox 1996). *Sharia* is the ideal for domestic governance. Good governance in this context refers to those initiatives and practices that advance the strength, coherence, and construction of Islamic civilization, ideally governed by a universal Caliphate with the coming of the Mahdi, and ruled by him. Political disintegration in the Middle East (the "Arab Spring" et al.) advance universal Islam, and it may be argued that, with the formation of the ISIS movement, here again we see a nostalgia for organization as an "axial empire" in the form of a universal Caliphate, and perhaps the longing for the Islamic messiah.

Thus, it could be argued that we apparently are witnessing a proliferation of anti-modernist impulses and actions around the world in recent years. Is the Western notion of modernity universally adaptable? Even in the West, we see a backlash against the most recent program of modernity, globalization; hence, it might be better to ask, whose modernity?

Populism, social atomization, and the emergence of strongmen: Populism and liberal vs. illiberal governance

As I glance around the world in 2017, it seems to me to be unfruitful to philosophize or speculate on the nature of modernity and its varying cultural manifestations without pausing to note and reflect upon the significance of the recent emergence of what we might call "strongmen" across regime types. Vladimir Putin seems to be the "late modernity" strongman prototype. He has taken a putative fledgling democracy and – through the force of a strong personality, a taste for power, and the capacity to build an oligarchical coalition – has converted it into a corporatist semi-authoritarian state that appears to function more like a syndicate than a state (Blank & Kim 2016). This is in some manner unsurprising, as Russia has no sustained history of democratic institutions and seems ever susceptible to the messianic pull of the *vozhd*. Meanwhile, Hugo Chávez in Venezuela exploited a fortunate oil resource endowment (petroleum once valued at over $100 per barrel) to attract a teeming population of young and urban poor who were happy to support leftist populism as long as the government handouts continued. By contrast, Chávez's successor is finding that political support is much harder to come by with oil at $50 per barrel oil and hyperinflation a daily battle (The Economist 2017). Voters in the Philippines have handed presidential power to Rodrigo Duterte, who fashions himself a latter-day Filipino "dirty Harry," a defender of the poor and defenseless and the scourge of maleficent drug dealers, upon whom his police and those in their pay have already visited some 7,000 extrajudicial killings (Tan 2017). Recep Tayyip Erdoğan has weathered a military coup attempt and continues to move to silence dissent with evident support from a burgeoning Islamic segment of the Turkish population (Ashdown 2017), while Hungary's Viktor Orbán "rails against the European Union, has built border fences to keep out refugees, and wants to create 'an illiberal new state built on national foundations'" (The Irish Times 2017).

Closer to home here in the "liberal" West, we have recently seen two stunning victories of populist movements at the very heart of what the left has always criticized as "Anglo-American neoliberalism." The British people almost shockingly have chosen to leave the European Union, and at this writing the court-ordered parliamentary confirmation vote of the popular plebiscite results has been held and Parliament has confirmed the popular vote's demand to leave. The Conservative government of Theresa May has chosen to come "all the way out" rather than try to arrange something like "Europe lite." The likely significant consequences for the British economy (including a plunge in the value of sterling to something like $1.20 US) so far have not arrived. Perhaps even more stunning has been the wholly unpredicted election of the populist firebrand Donald Trump to the Presidency of the United States on the Republican Party ticket.

These strongmen have come to power on waves of populist sentiment that are domestically very much akin to mass movements, if milder than the great historical mass movements. An adequate understanding of change in societal values and conventions must begin with an understanding of the social psychology fostering such change. Armed with that knowledge and thus trained to recognize the behavioral symptoms of such "mass societies," we can identify those shifts in societal values and conventions. The sociologist William Kornhauser (1959: 123) wrote that "people cannot be mobilized against the established order until they have been divorced from the prevailing codes and relations. Only then are they available for activist modes of intervention in the political process."

The availability of non-elites for mobilization is high when society has been atomized by various discontinuities. I will describe these shortly, but it is important to note the special vulnerability of both intellectuals and the lower classes to the siren song of populist and social movements. Intellectuals are vulnerable because "they more keenly experience the lack of larger purpose than do those less given to the abstract and symbolic" (Kornhauser 1959: 62). The lower classes, for their part, "are especially susceptible to the belief that all personal misfortunes are due to conspiracies against them, and that the world is divided into 'people who run things' and 'people like us.' They are readily attracted to ideologies that paint moral issues in rigid and absolutistic [*sic!*] terms" (Kornhauser 1959: 73). But while "unattached" intellectuals are often found in the vanguard of populist and mass movements, it is not always so. Even if they were, no populist movement could arise without a deeper base of support than that of unattached intellectuals.

> Any political enterprise which aspires to power on the basis of popular support will have to command the allegiance of sizable numbers of people from both the middle and working classes… the size of the classes makes it a necessity (Kornhauser 1959: 194).

Those from the middle classes most subject to the appeal of populist movements are its "marginal" members, including small businessmen and small farmers. They have little in the way of a common base for relations with other classes and tend to be go-it-alone people, resentful of larger competitors and fickle customers and clients. The modern shopkeeper or small businessman "is the truly marginalized man in industrial society" and thus available for mobilization by populist movements (Kornhauser 1959: 202). Thus, social and economic competition among marginal middle classes is a force for social atomization. Hannah Arendt (1973: 313) tells us that "the competitive and acquisitive society of the bourgeoisie had produced apathy and even hostility toward public life." Extremism among farmers is attributed to similar competitive pressures, as well as to their "isolation" (Kornhauser 1959: 207). Yet populist and mass movements are not predominantly a class phenomenon. This has been particularly not the case since the 20th century, when the concept of "the masses grew out of the fragments of a highly atomized society whose competitive structure and concomitant loneliness of the individual had been held in check only through membership in a class" (Arendt 1973: 317).

Kornhauser (1959: 159, 143, 129) argues that it is discontinuities in society, community, or authority that generate an atomized society, destroying social support networks and leaving individuals isolated and exposed. Discontinuities in authority can result from intended or unintended causes, and tend to disorient a society from its normative conceptions of legitimate social purpose. Intentional causes tend to be policies of authoritarian and totalitarian elites deliberately to keep society in a state of anxiety, insecurity, and political isolation. Such policies have relied on a variety of means: forced migration, terror, and purges, to name but a few. We do not see change in social conventions every time there is some slight communal, societal, or regime change in a given time and place. To assess the likelihood of social atomization from change and the attendant likelihood of change in conventions, one must examine "factors associated with major discontinuities in the social process as measured by the rate, and mode of social change" (Kornhauser 1959: 125). How quickly and dramatically change comes, as well as its differing modalities, are crucial factors in determining whether it will produce the social psychology of the populist or mass man in the observable behavior of the target populace.

Though *urbanization* is an important factor in producing social atomization, both its rate and modalities must be examined to draw conclusions about causal significance. Mass and populist characteristics of society are most often found when one can observe "especially rapid rates of changes in the size and composition [of the people] residing in an area" (Kornhauser 1959: 145). Similar caveats must be applied to both industrialization and urbanization as preconditions leading to social atomization. Clearly, one must look at the circumstances under which industrialization was introduced as well as the rate at which it transformed older modes of production. It is not enough simply to focus on the extent of industrialization. Societies may, and do, adjust to this process to obviate atomizing factors, but a great deal of strife can develop among elements who are highly vulnerable to the more negative forces of industrialization if they have no social or institutional means to seek relief from those forces. Importantly, *the same may be said of the process of de-industrialization,* leaving behind joblessness and cash-strapped cities that become, in the vernacular, "postindustrial holocausts."

It is specifically an economic calamity such as steep recession or depression that tends to destabilize a society or community. A steady level of poverty, however grinding, will not produce the same effects. Only when economic conditions are changing for the worse do the kind of destabilizing frustrations develop among individuals and classes within society that favor the rise of mass or populist conditions and behaviors. These individuals begin to feel "frustrated and insecure as they compare their lot with the one that has been held out to them as their legitimate condition" (Kornhauser 1959: 159).

What is crucial here is the change in social status resulting from a rapid change in societal economic conditions. If social arrangements are so disrupted by an economic or some other crisis that people feel that they have no hope for future improvement of their conditions, the tendencies toward political activism grow. Prolonged unemployment is particularly atomizing as it destroys existing social ties. The longer the individual lacks employment in some continuing occupation, the more divorced he becomes from society. This dynamic can foster a devastating sense of personal humiliation associated with his prolonged failure to find employment with which to provide for himself and his family and validate his sense of self-worth as a needed, productive member of his community. At this point, the activism of a populist movement meets his need. It gives him something to do, and a sense of purpose which compensates for his feelings of uselessness and helplessness. This provides an enormous appeal to the long-term unemployed on psychological as well as economic grounds (Kornhauser 1959: 162–66). I would argue that it rather goes without saying the *people who feel their community has been overrun by strangers who do not share their culture or their values also feel atomized and alienated by a discontinuity of their community of the first magnitude.* At the level of personal psychology,

> An individual who lacks opportunity for participation in society fails to receive support for a sense of his own worth and therefore finds it more difficult to sustain favorable attitudes toward himself. Self-estrangement in turn heightens the individual's readiness for activistic [sic!] 'solutions' to the anxiety accompanying personal alienation... For the individual who lacks a firm conception of himself and confidence in himself does not possess the basis for strong control over himself and is highly suggestible to appeals emanating from remote places (Kornhauser 1959: 107–08).

And thus:

> A characteristic response to estrangement from self is a diffuse anxiety and a search for substitute forms of integration. The alienated individual's lack of ego integration makes him susceptible to manipulation.... [T]he individual who is self-alienated is forced to turn to the mass opinion for directives on how to feel about himself (Kornhauser 1959: 109).

The consequences of social atomization for major national societies, their governance, and global governance

Among citizens of the Russian/Orthodox cultural region, the enormous discontinuities in the economy after the fall of the Soviet Union were obvious and devastating. The collapse of the command economy was followed quickly by the collapse of state-owned enterprises. The economic crisis in Russia after 1991 left people bereft of employment and without the means to provide for themselves and their families. Massive inflation wiped out the savings of almost everyone, a burden particularly devastating for the well-being of pensioners. Discontinuities in community followed with the privatization of housing and the rise of mafias, abetted by the decline of policing in the absence of state funds to pay the police. The average life expectancy of Russian men fell to as low as 54 years, in no small measure due to the ravages of rampant alcoholism that preyed upon men who had lost their positions, their incomes, their standing in society, and their nation. Discontinuities in society included an attenuated and – in parts of the former Soviet Union, such as the Baltic States and the Ukraine – an amputated citizenship. Rising crime rates and the evaporation of solidarity completed the assault on society, capped off by a dramatic reduction in status for many. Discontinuities in authority could not have been more vivid, with a complete collapse of central political authority resulting in a horrifically wounded Russian national pride.

Vladimir Putin's strong-armed state syndicalism has been the willing heir to collapsed Soviet power after a historically brief period of genuine parliamentarianism squandered by economic mismanagement, corruption, and election exhaustion. Putin's supporters explain why Putinism and Russian syndicalism continue to dominate governance structures in both Russia and its "near abroad." Explanations tend to take the form of assertions such as: Putin is a strong leader, Putin built the Russian middle class, Putin has improved social welfare, Putin has restored Russian might, Putin is restoring *Novorossiya*, Putin has a black belt and an iron handshake, and there is no one else. His supporters believe that he is the answer to and redeemer of all of Russia's discontinuities in economy, society, and community.

In Europe, the English-speaking countries – and even the United States in recent years – have seen discontinuities in their economies, trends that have had significant consequences for supranational governance and global liberalism. The result of these discontinuities has been broad popular dissatisfaction with wage stagnation and the waning of domestic manufacturing. These changes have been particularly difficult for working people, generating widespread income stagnation and economic insecurity. Discontinuities in community have featured a rural brain-drain and post-industrial holocaust cities in the rust-belt, particularly in the US, while Hispanic and African immigrant demographic movements have generated radically transformed neighborhoods in small towns as well as cities in which the original denizens often feel themselves to be under siege by the invasion of a foreign culture. Discontinuities in society, then, have displayed a tendency toward hyphenated citizenship, rising crime rates, ghettoization of immigrants, the relative slowness of cultural assimilation, and a reduction in social status. Discontinuities in authority in such places have sparked a stunning backlash against societal and political elites.

Most notable in recent months in this context has been the vote of Great Britain to leave the European Union (the "Brexit") and the election of populist candidate Donald Trump on the Republican ticket to the American Presidency. Shortly before the election, Trump supporters tended to explain their allegiance to the candidate in the following terms. I take the liberty of paraphrasing and summarizing: "He's not a politician" (anti-elitism). "He says what we're thinking" (populism). "He's a successful businessman" (economic insecurity). They cited Trump's campaign pledge to build a wall on the Mexican border and impose a ban on Muslim immigration (nativism). "He wants to make America great again" (nationalism). Trump in America, like Putin in Russia, is the answer to and presumptive redeemer of all of the discontinuities in economy, society, and community suffered by working class (and a substantial portion of middle class) Americans.

When British supporters of the Brexit explained their preference, they proposed pressing issues such as stagnant wage growth. They tended to be unemployed or underemployed. They were troubled by the pace of demographic change in recent years. Older Britons tended to vote to leave, younger students tended to vote to stay. The attainment of higher education and the possession of special skills tended to be a delimiter. "Haves" voted to stay, while "have nots" voted to leave.

Supporters of the Brexit tended to insist adamantly that the referendum was a question of sovereignty rather than economics. Brexiters even could be found in significant numbers among ethnic minorities in Britain, long-time immigrants whose rationale was concern with recent immigrants: "We assimilated culturally, but they won't." Disaffection from and even distaste for the European Union was a persistent rationale for voting for Brexit, with the assertion that the EU is and has always been an elite-driven project with a huge democratic deficit. Discontinuities in society and community, rather than economy, dominated the rationale for the Brexit vote. Britons have made it clear they are willing to troll in whatever turbulent economic waters they must. They want control over their borders, their laws, and their demographics.

Sources of social atomization are also not difficult to locate in the Confucian world, particularly in China, due to its economic successes and excesses of the past four decades. Discontinuities in economy include the significant challenges the country has in maintaining rapid economic growth. There is strong evidence of massive over-investment, particularly in property. Some observers estimate that there is a backlog of unsold housing that is the equivalent of five New York Cities of flats, even as housing prices in Tier 1 cities exceed the means of the average wage earner by many times. Particularly problematic here is dependence on credit-fueled economic growth, migrant labor wages, employment challenges, and – by no means least – an upcoming demographic cliff that, under present trends and projections, will see 600,000,000 fewer Chinese by 2045 as an ageing population reaches its life expectancy. Discontinuities in community include the struggles by many Chinese to obtain access to housing, schooling, and medical care. Furthermore, the restrictions of the *hokau* residential permit system make life difficult for migrant laborers. The supreme irony is that the men who built the Shanghai skylines cannot live there with their families or obtain schooling or medical care for their children. The burden of caring for ageing families will significantly challenge young people as the "one child" policy has left many young couples responsible for financing the retirements of two sets of parents and four sets of grandparents, while shouldering the burden of raising their own children. Discontinuities in Chinese society are persistent, and feature a high Gini coefficient (of economic inequality), just below that of India and even greater than that of the United States. For some time to come, these discontinuities will demand some sort of settlement between the urban privileged and migrant and rural underprivileged who have seen less of, or even little of, the economic prosperity that the Deng reforms and the subsequent four decades of economic development have brought to China. A stubborn systemic tendency toward corruption also persists, in spite of a long and spirited campaign by Xi's government to uproot it. If allowed to entrench itself further, the threat it poses to the continued economic development of China could be mortal. Ironically, continuities in authority also could be rising due to a dearth of outlets for the public to express its complaints and dissatisfactions as well as an apparent demand for absolute loyalty to the Party by the government. Observers tend to disagree on the question of whether the apparent unyielding unwillingness of the government to accompany vast economic reform with even a modicum of political reform is a conservative policy or a quite risky one.

China's governance remains a work in progress. The question of the future of global governance is similarly unclear. The blows to European supra-nationalism and the threat of rampant populism (particularly in the West) to a global trading order dominated by states with cosmopolitan citizenship bode ill for a global governance still committed to the prescriptions of Kantian Enlightenment rationalism and idealism.

PART II

Development

3

ALTERNATIVE MODERNITIES ON THE ROAD TO NOWHERE

Jack Snyder

A naive version of liberal modernization theory, existing especially in the minds of its critics, anticipated that sooner or later all societies would modernize and wind up looking like the trailblazer on that path, liberal England. As Daniel Lerner put it, "what the West is..., the Middle East seeks to become" (Lerner 1958: 47; cf. Gilman 2003). A number of the most interesting modernization theorists, however, wrote about the detours that authoritarian modernizing countries like Germany, Russia, and Japan had taken off the main road (Moore 1966; Gerschenkron 1962; Huntington 1968). For the most part, these scholars saw such alternative modernities as ephemeral successes, leading ultimately to disastrous dead ends.

By the end of the Cold War, liberal triumphalists thought that such alternatives had been relegated to history (Fukuyama 1989), but now that judgment seems premature. The rise of China, the impressive achievements of Singapore (Acemoglu & Robinson 2006: 8–10), the dynamism of politicized religion, and the assertiveness of authoritarian regional powers have lent credence to the notion that there might be multiple ways to be successfully modern, including illiberal ones (Eisenstadt 2000; Katzenstein 2010). Liberalism's recent woes – the 2008 financial crisis, the economic slowdowns in Europe and Japan, Europe's failure to integrate migrants, and endemic terrorism – have stoked support for illiberal, populist alternatives even in core liberal states. Some commentators perceive benefits in the growing diversity of powerful states and dynamic cultures in world politics (Acharya 2014). Especially in societies that lack the embedded cultural and institutional legacies that allowed liberalism to thrive in the West, forms of modernity that foster technical rationality while tempering liberalism's free-wheeling competition in politics and markets might work better to sustain stability and growth. Pragmatic illiberal powers constructively might encourage dogmatically liberal stewards of international institutions to adopt more flexible practices that would tame the creative destructions of unregulated capitalism and smooth the clash of civilizations through live-and-let-live accommodations. If so, a multiplicity of alternative modernities could live alongside the liberal form, much as European state-organized capitalism coexisted with Anglo-American laissez faire (Hall & Soskice 2001a). Cultures of rights could find a language of dialogue with cultures of duties. Even the quintessential liberal philosopher John Rawls (1999: 27) endorsed comity among "decent," even if illiberal, states in the international system. This is said to be "the ASEAN Way" (Acharya 2001b).

In contrast, other analysts accept the inevitability of multiple modernities, including illiberal ones, but see this as a source of breakdown of the system of liberal rules that sustains global order and prosperity. Dynamic illiberal powers may be profiting from the open international economic order that liberalism created and sustains, but their instinct may be to ride free on global public goods provided by liberal institutions, jockey for advantage in ways that undermine liberal rules, and defy the principles of law and individual rights that are the ultimate foundations of that system. In this view, the diverse collection of authoritarian technocracies, clientelistic petro-states, religion-based regimes, and populist semi-democracies will be unlikely to converge on any new set of rules and institutions to replace or even constructively amend the liberal order. If so, this will be "No One's World," managed by the "G-Zero" (Bremmer 2012; Kupchan 2012).

The conjecture that I advance differs from both of these views. Instead of anticipating multiple, durable forms of modern social order, I speculate that we are witnessing convergence on a single main competitor to the liberal order: namely, populist nationalism, which – in a contemporary form of the problem analyzed by Karl Polanyi (1944) – is an attempt to reconcile the currently dominant form of inadequately regulated capitalism with heightened mass public demands organized at the national level (cf. Caporaso & Tarrow 2009; Ruggie 1982; Berman 2006: conclusion; Blyth 2002: 1–26; Stiglitz 2003: chs. 1, 4). To be sure, this competitor to liberalism takes on different forms in different settings, depending on whether the country is an advanced democracy or a developing state, and whether the lines of political cleavage and faction are ethnic, religious, nativist, or class-based. But the underlying logic is similar, gravitating toward similar governance formulas and legitimating ideologies.

Its taproot is the disruptive social and political impact of market forces unleashed in the course of economic modernization, both at the level of the individual society and the global system as a whole. The increased mobility of capital and people that is called globalization has created opportunities for economic growth, but also huge challenges to adapt political and economic institutions to this altered world. Many of the elites who are riding this tiger of change have been seeking to fashion institutions and ideologies that allow them to maintain power and privilege in the face of rising demands from both the winners and losers from globalization. Populist nationalism is the formula that many of them are turning to, with considerable tactical success, toward this end.

My argument differs not only in seeing a single competitor to liberalism, rather than multiple ones, but also in seeing that competing formula as unsustainable and doomed to failure. Market-taming authoritarian populism was the formula that animated most of the failed historical projects trying to reconcile authoritarian, elite-managed systems with technically advanced modernity in the face of rising mass activism spurred by economic change. Usually this took the form of nationalism, but in Soviet Russia a functionally comparable idea was called "socialism in one country" (Carr 1958–64).

For reasons embedded in the normal sequence of economic development, this formula can be wildly successful in early stages of the shift to modern market society. But it fails at the stage where what Karl Marx (1992: section 8, chapter 26) called the "primitive accumulation of capital," by tapping a large pool of underutilized agricultural labor, must shift to productivity-based growth. These two stages of development call for different institutional arrangements, indeed different economic cultures, but at the inexorable turning point, elites often choose to maintain state-organized markets justified by populist, usually nationalist, ideologies just when they should be choosing impersonal rule of law and democratic accountability instead (Huang 2008). These efforts tend to fail dramatically in the long run, but not immediately, so elites often gamble on the sustainability of this formula, at huge cost.

I analyze the details of this mechanism by drawing on research on "late development" and what has recently been labeled "the middle-income trap" (Gerschenkron 1962, ch. 1; Dollar 2015; Eichengreen et al. 2013; Snyder 2017).

The logic of modernity

In considering the feasibility of multiple pathways to modernity, it is important not to smuggle in liberal assumptions by definition. For that reason, I define modernity in a minimal way as a social order that produces self-sustaining economic growth based on scientific and technological progress. I leave out of the definition any assumptions about the institutional features that are needed for a modern social order, since that is what we are trying to figure out. I also leave as an empirical question whether the relevant unit of analysis for a given social order should be considered the nation-state, an economically interdependent multinational region, or the globe as a whole.

Modernization theorists since Ferdinand Tönnies' breakthrough work on "community and society," *Gemeinschaft und Gesellschaft* (1887), have posited that the crucial move enabling modernity is the shift from personalistic social relations based on family, lineage, patron–client networks, and cultural in-group favoritism to impersonal social relations based on rules that apply to all individuals. Similarly, Emile Durkheim's *Division of Labor in Society* (1893) posited a transition between two distinct forms of social solidarity, from traditional society's group solidarity based on similarity to modern society's solidarity based on complementarity of functional roles in its complex division of labor. Durkheim went so far as to claim that the whole idea of the individual and of individualism emerged from this change in social organization. Other foundational figures of social science filled in other pieces of the modernization picture. Karl Marx already had analyzed the breakdown of feudal caste privileges as ushering in capitalist relations of production based on free contracting. Max Weber in turn discussed the shift from organized nepotism to rational, legal, meritocratic, rule-following bureaucracies.

In this framework, culture matters as well as institutions. But the crucial cultural divide is not between civilizations bearing different cultural legacies, as in the theory of multiple modernities, but rather between the culture of tradition and the culture of modernity (Katzenstein 2010; Eisenstadt 2000). When human rights scholar Jack Donnelly (2003: 107–26) collected a list of purported "Asian values," it turned out they are not about cultural distinctiveness at all, but actually express either values typical of traditional societies everywhere, including the historical West – e.g., patriarchy, duties rather than rights, prioritizing society over individuals – or typical concerns of all developing countries, such as asserting state sovereignty and prioritizing economic development over civil rights (cf. Krasner 1985).

These ideas not only animated the American modernization theories of the 1950s and 1960s, but they remain central to contemporary social science works on the evolution and efficacy of the modern state. Francis Fukuyama's two-volume, 1,250-page masterwork (2011, 2014) argues convincingly that the central problem of political order and decay from prehistory to the present has been the struggle to overcome the inefficiencies embedded in lineage-based, clientelistic social systems, supplanting them with modern systems of impersonal rules and accountable government (cf. Fukuyama 2015; North et al. 2009; Acemoglu & Robinson 2012). Celibacy, eunuchs, and orphan slave armies were just some of the idiosyncratic institutional innovations devised to overcome the deadweight costs of nepotistic corruption in earlier times, until impersonal rule-of-law institutions eventually removed the fetters on growth in liberal societies. Ending on a somber note, however, Fukuyama argues that remnants of the old corruption remain endemic in institutional legacies even in the most advanced democracies, not to mention rising state-dominated capitalist powers such as China, with dangerous implications not only for inefficiency but also disorder.

A crucial question for the debate over multiple modernities concerns how far a society must go in adopting the full package of liberal social arrangements in order to achieve self-sustaining economic growth. The purist view emphasizes that virtually all societies that have been highly successful over a long period of economic development have moved quite far in the direction of the fully liberal model, both in formal arrangements and in effective rights for most segments of society. This includes due process of law, non-discrimination, rule-based protections of property and sanctity of contracts, and widespread rights to political participation through free speech, political organizing, and fair, competitive elections of representatives that are bound by law. Setting aside oil sheikhdoms and city-state entrepôts that exist under the protection of their liberal customers, the correlation between per capita income and stable, liberal democracy remains overwhelming (Przeworski et al. 2000; Haggard & Kaufman 2016: 129–30).

That said, China's unprecedented run of sustained economic growth raises the question of whether its illiberal formula, based largely on the technocratic skill of its elite, can succeed indefinitely (Bell 2015 argues in the affirmative). Modernization theory might pose it this way: Is Weberian technical and administrative rationality enough to sustain modern economic performance, especially if the system's legal rationality is poorly developed? Below I will answer this question skeptically, based on the literature on typical sequences in the pattern of late development.

Other unanswered questions in modernization theory challenge the assumption that liberalism even in developed democracies is safely past the danger point as a durable solution to sustain economic growth and stable social order. Three contradictions within contemporary liberalism stand out, most of them exacerbated by the accelerating globalization of the division of labor.

First is the tension between liberalism's two core concepts, liberty and equality. Since Marx, critics of liberalism have argued that the formalism of equality in legal rights is a practical dead letter if unfettered pursuit of grossly unequal wealth produces grossly unequal political influence. Although Rawls may have solved this problem in theory, the increased mobility of global capital has exacerbated it in practice.

Second is the tension between equal civic rights within the nation-state and huge inequalities of rights, opportunities, and outcomes among individuals in different nations, notwithstanding their joint participation in single, global division of labor. In earlier times, when modernization theory could assume limited economic interdependence across national boundaries, such inequalities were arguably of secondary concern, but now they are posing functional problems in governing the increasingly mobile international financial system and labor markets (Chinn & Frieden 2011: 171–74).

Third, unregulated markets produce innovation and growth through a process that Joseph Schumpeter called "creative destruction." As Karl Polanyi (1944) argued, however, politically engaged modern mass publics tend to insist on state protection from the wrenching pain of market adjustments. They are likely to support Nazis, Communists, economic nationalists, nativists, Keynesians, welfare-state corporatists, or anyone who promises to use state power to control the invisible hand of the market. This creates an interlocking set of endemic troubles for liberal modernity. Market "self-corrections" can careen out of control, as in 1929 and nearly in 2008, bringing the economic system down (Frieden 2016; Drezner 2014). In parallel, public demands for protection from this danger can play into the hands of illiberal demagogues, bringing the political system down. Finally, owners of capital can use ideology, market power, and political leverage to resist demands for market regulations that limit capital mobility and profitable risk-taking. During the Kennedy Administration, Walter Heller and the Harvard school of liberal economists thought that the Keynesian consensus and Bretton Woods institutions had solved such interlocking problems by embedding global and national market regulation in a set of prudent rules, but globalization and deregulation have undermined these stabilizers (Blyth 2002: 95, 152–201).

Today's newly rising powers confront all of these contradictions, as well as another set of even worse contradictions that are distinctly their own. Since all of the rising powers are extensively integrated into world markets run according to liberal rules, they face the ripple effects of the same contradictions that the established liberal states do. All of their domestic economic and political systems, notwithstanding areas of authoritative state control, are penetrated by the practices, ideas, and dependencies of global markets. But in addition, the rising, modernizing powers also face an even more challenging set of contradictions that are specific to their own transitional circumstances. These are rooted in the mismatch between the institutions these emerging powers have and the ones they need for further growth.

In the next section, I discuss the political implications of the contradictions that advanced democracies and middle-income states face in a globalized era, explaining the appearance of similar populist, nationalist forces in both. In a subsequent section, I discuss the distinctive additional contradictions that make for a particularly potent version of this brew in today's rising powers, the BRICS and near-BRICS.

Populist nationalism under globalization

The syndrome of nationalist or nativist, culturally conservative, populist authoritarianism has been riding a rising tide both in middle-income developing powers and also in the core regions of the developed democracies, the EU and the US. Not only are the regimes of Putin, Erdogan, and Modi often said to fit this profile, but so do the constituencies for Brexit and Trump (de Bellaigue 2016). Far right European parties manifesting such features have participated in coalition governments in Austria, Croatia, Estonia, Finland, Italy, Latvia, the Netherlands, Poland, Serbia, Slovakia, and Switzerland, and they have supported minority governments in Bulgaria, Denmark, the Netherlands, and Norway. They have also been influential in the politics of France, Belgium, and Hungary (Golder 2016: 478). In some prominent cases, far right movements exist in a symbiotic, if sometimes fraught, relationship to mainstream market-oriented conservative nationalist parties: the Jobbik party and Prime Minister Viktor Orbán in Hungary, Israeli settlers and the Likud, the Tea Party and the Republican Party in the United States, and the RSS and Prime Minister Narendra Modi in India (Aron 2016).

Consequently, any explanation for the rise of illiberal populist regimes and movements in the rising major powers also must be able to account for analogous regimes and movements in the developed democracies. My conjecture is that the trend in the rising powers is an especially thorny form of a more general phenomenon that can be best illuminated by looking first at established liberal states.

Populism, like nationalism, is often said to be an empty signifier that can be filled with almost any content and its opposite, and adapted for almost any constituency. It can be on the left or on the right; militantly religious or militantly secular; prone to violence or, like William Jennings Bryan, militantly pacifist. That fluidity allows populism and nationalism to be successfully pressed into the service of seemingly improbable masters. Wheeler-dealer businessmen like Donald Trump and Thailand's former Prime Minister Thaksin Shinawatra can win votes as populist celebrities promising to deliver their constituents from the brutality of market competition. Plebiscitary dictators from Juan Peron to Vladimir Putin can preside over "sovereign democracies," through which the people are promised they exercise national self-determination without actually having procedurally accountable government. These protean characteristics help explain how Herbert Kitschelt, the eminent scholar of far right European parties, could argue in 1997 that their natural winning ideological formula is a pro-market position on the economic dimension with an authoritarian position on the cultural dimension, whereas nowadays anti-system parties are more likely to "blur their actual economic position in an attempt to maintain a cross-class coalition" (Golder 2016: 490, citing Kitschelt 1997; cf. also Rovny 2013).

While many populist parties have been on the right, some espousing pro-market ideologies, some are on the left, lashing back against the effect of globalized markets. Nonetheless, they often share common attitudes on other issues. A 2016 survey of 45 "insurgent parties" in Europe, "ranging from the hard left to the far right," revealed a common skepticism about the EU, uniform backing for holding popular referenda, wariness about relations with the US, prioritizing refugees and terrorism as a greater threat than Russia, distancing from Ukraine, and lack of enthusiasm for sanctions against Putin's regime (Dennison & Pardis 2016).

If the policy content of populism can be varied and elusive, students of these movements argue there is nonetheless coherence to its political style. Populists, like nationalists, portray a mythologized communitarian version of "the people" as the repository of virtue, their sufferings the result of a parasitical, immoral elite (Golder 2016: 479; Moffitt 2016). Consistent with a myth of the general will of the people, populists believe in a form of democracy that is majoritarian and plebiscitary. The people's will is unitary, not pluralistic, and is anchored in the traditionally dominant religious, ethnic, or racial majority group of the nation, which fears that it is being sold out by its own elite. The popular will is seen as best expressed by referendum in a direct relationship between the people and their leader, unmediated by laws and institutions that only introduce stumbling blocks in the path of implementing the popular will by decisive means. Populism is inclusionary to a fault within the in-group, with little tolerance for individual deviations that sully its purity, and exclusionary toward out-groups, which are at odds with the myth of popular unity. Like other mass social movements, populist movements are not well suited ideologically or organizationally for compromise (Kitschelt 2003: 84). They see a world in crisis and favor "bad manners" as a way of defying more genteel elites and polite mainstream discourses that obfuscate the urgency for action (Moffitt 2016: 44–45). Even when populists are actually in the minority of public opinion, their imaginary hoped-for world is a tyranny of the majority.

Social scientists' explanations for the rise of far right populism look both at the demand for populist ideas among the people and at the supply of facilitating conditions in the social environment. On the demand side, a recent survey of this research identifies three interrelated sources of complaint that feed far right movements: modernization, economic grievances, and cultural grievances. The first focuses on the losers from modernization who struggle to adapt to the wrenching social and psychological effects of changes driven by technology, deindustrialization, globalization, democratization, the transition from socialist to capitalist society, and the importation of values from the US and Western Europe. European far right voters tend to be young males with low levels of education and insecure employment from "the second-to-last fifth of postmodern society," often motivated by pragmatic concerns (Golder 2016: 483; Minkenberg 2000: 187; van der Brug et al. 2005). This overlaps with economic grievance inasmuch as far right voters in Europe are those most likely to believe that they are competing economically with immigrants. However, such perceptions depend on how politics structures economic competition. In Britain, for example, nativist rivalry with immigrants happens in locations where economic scarcity combines with immigrant voting power to shift public resources away from the native population, triggering resentment (Dancygier 2010). Cultural grievance against immigrants is a universal theme among Europe's far right parties, though voters for mainstream parties often express anti-immigrant attitudes, too. The size of the immigrant population does not seem to correlate in a straightforward way with the intensity of the cultural grievance.

On the supply side, research focuses on how the structure of political competition creates openings for populist parties and movements, or fails to do so. A central concept is the political cleavage structure. Historically, for example, the main axis shaping European party systems has been economic class cleavage. When this cleavage structure was stable, there was little space for populist movements to gain a foothold. Some argue that class cleavage has become less important in the postindustrial social welfare states of Western Europe. Others note that the party systems and economic class structures in Europe's post-communist states are fluid and ill-defined. For these structural reasons, far right parties gain more room to maneuver by redefining grievances, including economic ones, in cultural terms (Inglehart 1977; Kitschelt 1988, 1997; Evans 2005; Golder 2016).

Viewed from the perspective of privileged elites, democratic systems inherently create a risk that the median voter will want to use steeply progressive taxation to equalize wealth (Acemoglu & Robinson 2012; Boix 2015). To what extent this actually occurs is quite variable from country to country and over time. Equality of wealth was greatest in advanced capitalist states in the wake of high taxation to pay for the world wars. More recently, inequality has been growing along with globalization and the increasing adoption of market-fundamentalist policies, but at different national rates (Piketty 2014; Scheve & Stasavage 2012).

One tried and true method that elites use to escape confiscatory taxation in democracies is to attempt to shift the main axis of political cleavage from economics to group identity, nationalism, and culture. In the United States, this dynamic was colorfully captured in journalist Thomas Frank's book, *What's the Matter with Kansas?* (2004), which argued that wealthy Republicans were duping low-income, small-town, white citizens into voting against their own economic interests by hyping cultural "wedge issues" such as abortion, gay rights, racism, and threat-inflated militaristic patriotism (cf. Bartels 2005).

Variants of this tactic are ubiquitous across time and space. In the wake of Germany's rapid industrialization, the aristocratic monarchist Chancellor von Bismarck responded to middle-class demands for constitutional government and limited parliamentary democracy by going them one better: universal manhood suffrage including the working class and the peasantry. He gambled that the peasants would vote the way conservative landlords told them to, and that appeals to true German national identity could split up any hypothetical progressive coalition among labor, Catholics, and middle-class Protestant nationalists. He was right. In nine national parliamentary elections between 1870 and 1914, the conservative coalition did much better in the five that were fought on so-called "national" issues defined by the Kulturkampf against the Catholics, colonial expansion, and military budgets to defend Germany against "hostile encirclement" by the Entente powers, which German diplomacy had largely provoked in the first place (Fairbairn 1997: 48).

The all-purpose logic of shifting the axis of electoral politics from economics to identity politics also explains urban rioting in India. Steven Wilkinson's definitive research (2004) shows that riots occur when municipal elections are expected to be close between an elite-dominated identity-based party, such as the Hindu nationalist BJP, and a lower-class-based party attempting to appeal across identities such as Hindu and Muslim. Using thugs and rumors to foment rioting is designed to polarize politics on the identity axis on the eve of the election, inducing lower-class Hindus, for example, to vote their blood rather than their pocketbook. This strategy only works, says Wilkinson, when the state's governing party coalition, which controls whether state police will intervene to prevent the riot, does not include the identity group to be targeted in the rioting, such as Muslims.

Although supply and demand for such strategies is hardly unique to contemporary times, the current moment of capital mobility and demographic change facilitates their use. The next section traces these effects in the rising powers of the developing world.

Populist nationalism in emerging powers undergoing social transition

Over the past two or more decades, several of the large states in the developing and post-communist worlds have experienced significant economic growth as they introduced liberalizing reforms in their domestic markets and international economic relations. Impressive improvements in mass living standards and spikes in economic inequality have often accompanied this growth. I will lay out the case that a syndrome of nationalist, populist authoritarianism reflects economic and political contradictions that are heightened at the cusp of a shift from extensive to intensive growth strategies.

During the Cold War, large developing states for a time were able to pursue strategies of state-led economic development and import-substituting industrialization, or ISI (O'Donnell 1988). This depended economically on the "advantages of backwardness": mobilizing underutilized labor and resource inputs, copying well-known industrial processes, and using state power to accumulate capital and protect infant industries from foreign competition. Politically, it typically depended on authoritarian or single-party politics, with the state aligned with a coalition of domestic manufacturers and organized labor. Subsidies and protectionism were a drag on productivity growth, however, so the system gradually stalled out.

Around the same time, "neoliberal" policies of easier capital mobility and deregulation of markets, increasingly favored by advanced capitalist states and international financial institutions, offered opportunities for more dynamic, export-led growth. China, India, Brazil, Turkey, and other large developing states signed up for liberalization, and as a result each for a time experienced an acceleration of economic growth (Mukherji 2007; Huang 2008; Akca et al. 2013). Russia went through its more superficial, petro-state version of market reform.

In many of these states, the initial phase of liberalization remained based on some of the same advantages of backwardness as in ISI, only the market now became global. Cheap labor, easily extracted natural resources, copycat technology, and development-pushing state policies were now harnessed to foreign direct investment and integration into internationally managed production processes. The liberal institutional package of modernity was selectively cherry-picked, assimilating features needed narrowly for liberalized trade and finance, but broader rule of law, freedom of speech, and open democratic political competition were approached more warily. Clientelistic economic and political arrangements more commonly associated with traditional societies and state-run economies continued. "Neoliberal" in many such states became an epithet meaning crony capitalism.

Just as ISI eventually ran out of gas, so too there are strong signs that this phase of the neoliberal development model is leading into a transition trap at the point where extensive must give over to intensive growth. Extensive growth is based on adding more inputs: more labor coming off the farm, more land foreclosures, higher rates of capital investment, exploiting already known technologies on a wider market scale. But the very success of this phase leads to its demise. Wage rates rise once most of the useful labor force is employed. New land for commercial enterprises is harder to come by, and requisitioning it creates more resistance. The levers of state power ratchet the rates of savings and investment higher and higher, but force-feeding growth in this way demands more and more capital inputs to generate less and less output. This syndrome constitutes exactly the impasse that China faces now.

The solution is to shift from the strategy of extensive growth to that of intensive growth driven by increases in the combined productivity of all factors of production. The latter depends not on ever-increasing inputs but on the more efficient allocation of factors of production through responsiveness to market incentives. In countries in China's situation, this shift would require the development of its vast internal consumer market, which is at odds with the strategy of enforced savings (cf. Estlund 2017; Lee & Li 2014). But the major underlying requirement is for the strengthening of liberal institutions of rule of law and governmental accountability to reduce the inefficiency drag of corruption, insecure rights of property and contracting, and inequality (Shambaugh 2016).

This conclusion is well supported by research on the so-called middle-income trap among developing countries. Some economists have argued that middle-income countries necessarily tend to experience slowdowns once easy sources of growth such as rural-to-urban migration, primitive capital accumulation, and the initial spurt of export expansion are exhausted, while the institutional capacity for high-end growth through technological innovation and product differentiation remains underdeveloped (Gill & Kharas 2015). Others, however, challenge the notion of a middle-income trap, pointing out that episodes of slower-than-average growth can happen at every level and stage of development, and that high-income countries tend to have the slowest average growth. Even some who accept that the key to sustained growth lies in better quality institutions, especially rule of law and governmental accountability, suggest that institutional quality matters at all stages.

While taking these qualifications into account, a study by Brookings economist David Dollar (2015), finds that improving institutional quality is especially important for sustaining economic growth in middle-income countries, for precisely the reason of supporting the shift from extensive to intensive growth. China and Vietnam were able to develop relatively good institutions for their lower level of income and stage of development, and they sustained a good rate of growth for a time without developing the whole panoply of civil liberties. However, for high-income countries, he finds a tight connection between good economic institutions – including well-defined property rights, rule of law, effective government, and limits on corruption – and a broader set of liberal rights as measured by Freedom House's Civil Liberties index. Not counting oil states, Singapore is the only exception to this rule. In the 1990–2010 period, for countries at low levels of per capita income, authoritarian countries grew faster than democratic ones, but above one-fourth of US per capita income democracies grew faster. The developmental histories of Korea and Taiwan illustrate the pattern.

Dollar reports that by 2010 China and Vietnam no longer had above-average institutional quality for their income level. China fell well below average, and Vietnam fell to the average for its reference group. He notes that the subsequent growth slowdowns in both countries are consistent with his overall argument.

Putting this kind of analysis in perspective, the development economist Robert Wade (2016), long an advocate for state-led industrial policy in settings such as South Korea, accepts that the transition from extensive to intensive growth is likely to produce a slowdown. He argues, however, that fixing the problem not only requires improving legal and accountability institutions to prevent market failures; state investment policies also are needed to avoid debt traps and to boost the developing country's export profile into more sophisticated, diversified products.

If Dollar is at least partly right about the need continually to improve – meaning liberalize – economic institutions as a country moves up the per capita income food chain, why don't more countries, especially large ones, follow the example of South Korea and Taiwan? Several factors may come into play. Large size and substantial capital accumulated in the extensive growth phase give them leeway to pursue their preferred strategy. Moreover, administrative methods and patronage-based bargaining is what they know how to do, and they have succeeded at it. Like

Wade, they may think that industrial policy, done their way, will work to fix or anticipate market failures, as it has in their recent past (Naughton 2014).

A more basic reason is that the ruling elite, state apparatus, patronage networks, ethnic or religious majorities, and rising middle classes of these emerging powers have vested interests in keeping the incompletely liberalized system going (Hellman 1998; Nathan 2016; Teets 2014). The alternative is to hand political and economic power to the mass of the population, which in many such countries has not yet experienced the full benefits of market-based growth and indeed may have stockpiled grievances against crony capitalism. When the demonstrators in Tahrir Square learned that Hosni Mubarak and his sons would just get a slap on the wrist rather than a death sentence, the crime that focused their outrage was the ill-gotten vacation villas, not the repression of human rights activists.

To keep the game going, and to avoid reforms that would endanger their power and privileges, the ruling elite and its key support constituencies often turn to policies of authoritarian nationalism and cultural conservatism. Let us examine the elements of this syndrome one-by-one to consider the role it plays in this scheme of rule.

Plugging into the global capitalist system is the only game in town for the ruling elites. Soviet-style semi-autarky is a dead letter. Even continent-sized emerging powers now realize that their scale is not big enough to delink from world markets. Russia's economy depends on a diverse, developed-country market for its energy exports, not on the Eurasian Economic Union. China can build a Silk Road through Central Asia to burn excess construction capacity, but not to move to a higher level of per capita income and world-class innovative technology. Such states must participate in the global market system. The problem is how to package contemporary global capitalism and its attendant wealth inequality to citizens and subjects who do not see themselves benefiting much from it.

Authoritarianism is arguably the first choice for solving that problem (Diamond et al. 2016). But in a globalized market economy that depends on having an educated population with enough information and initiative to do their jobs, some pressure for accountability is inevitable. Even China wants to have local competitive elections and investigative reporting of local officials' abuses so that the central government can blame someone lower down. In Turkey, the Anatolian mindset of the tyranny of the new Islamic majority fits perfectly with Erdogan's view of democracy as "a bus you get off when you arrive at your destination," and the coup-plotters just drove it into the terminal. Even so, Erdogan sustains his rule not only through the repressive use of state power, but because he occupies the pivot of Turkish politics. The pious Turkish majority, including its thriving "Anatolian tiger" exporters, gives him a substantial base of support, but what really cements his party's rule is that his opponents prefer a coalition with him to alliances with each other: Kurds (most of whom are religious), secular illiberal hyper-nationalists who hate the Kurds, Gezi Park Western-style liberals, and the remnants of the Kemalist secular party burdened by its legacy of coup-plotting and elitist contempt for the people. In short, for most of these countries, authoritarian coercion is in the tool kit, but other political and ideological tools supplement it.

In poor countries with conspicuous wealth inequality, trickle-down economics can be used to justify economic inequality, as it is in developed states, and with a more convincing answer to the classic question asked in American political campaigns: Members of China's growing middle class, though it is still very small compared to the mass of the population, *are* much better off than their parents were (Kharas & Gertz 2010; Tsai 2007). Erdogan, likewise, is busting the national budget with no-bid mega-construction projects, digging new canals to parallel the Bosporus, and also dispersing construction patronage throughout the backwaters of Anatolia to build poor-quality universities and shiny new airports. This cements the loyalty of the support base, and to the gullible it looks like modernity (Ozturk 2016).

If these inducements lose their appeal, nationalism is a tested tool for shifting the main axis of politics away from economic interests and grievances to concern about culture and identity. It allows unaccountable or semi-accountable elites to pretend plausibly to be one with the people.

In rising powers, pride in the nation's newfound economic or political success gives an automatic boost to the regime. Putin has alternated economic and political prestige strategies, riding the invasion of Chechnya to power in his first landslide election, rising still higher along with the price of oil and gas, and now manufacturing prestige with his gambits in Ukraine and, more ambivalently, Syria. Along with the national pride that accompanies a booming economy comes the expectation that this newfound power should be accompanied by a newfound respect abroad. The expectation among Chinese nationalists that their century and a half of humiliation finally would be over was disappointed, however, in recent clashes over island sovereignty, leaving them upset not only with the non-compliant foreigners but also risking dissatisfaction over their own state's failure to deliver better results (Weiss 2013; Johnston 2013, 2017).

Another goad to nationalism is the risk that market exchange will be a vehicle for foreign penetration of the nation's traditional culture. Putin plays this "danger" to his advantage since it creates an excuse to shut down the homegrown progressive opposition and to urge vigilance against the decadence of the West. In Turkey, surveys show that the vast majority of Erdogan's supporters believe that the Gezi Park demonstrations were an imperialist plot by the West (Konda Gezi Report 2014). Erdogan, however, is safe from the charge that his market reforms created this Fifth Column, since the Kemalist regime that Erdogan supplanted had long ago empowered Istanbul's Europeanized "white Turks." All of the BRICS and near-BRICS, even those never subject to colonial domination, have well-rehearsed resentments stemming from the legacy of imperialism. Stalin's 1931 reminder that "Old Russia was always beaten for her backwardness" still plays fresh, as does the post-World War I Sevres Treaty for the Turks and the Opium War for China (Gries 2004: chs. 1–3).

With national wealth and power on the rise, constituencies who consider themselves the core of the nation, the sons of the soil, may expect it is time to be given their due economic rewards and cultural hegemony. Modi indulged Hindu nationalists by allowing the Gujarat anti-Muslim riots in 2002, and critics charge that he is similarly passive now when Christians are persecuted for beef-eating or trumped up charges of cow desecration.

In short, emerging authoritarian powers that want to shift the national discourse from the reform demands and economic grievances of those left behind to the cultural agenda of their support base will find plenty of opportunities to do so.

Assuming there is indeed a syndrome of authoritarian nationalist populism that has currency in a number of rising powers, the question arises whether this commonality will lead them to band together in rhetoric or in action. At the level of talk and symbolic action, the answer is already yes. The BRICS and near-BRICS are generally sovereignty hawks, balking at support for liberal humanitarian interventionism and decrying support for color revolutions. Ideologically, they support each other's justifications for majoritarian or "sovereign" democracy, cracking down on NGOs, and resistance to liberal rule of law, human rights accountability, and media freedom. They form their own international organizations and election monitoring organizations (Cooley 2017).

Nonetheless, alliances among authoritarians and ruthless nationalists have always been opportunistic. There has never been anything comparable among authoritarians to the principled democratic peace among liberal states. The trouble their nationalisms make for the liberal order may add up piecemeal, but it is unlikely to be highly coordinated. States like Russia and China will fear each other's power. That said, they will fear liberalism's principles as well as its material power, giving them an endemic motive for wary, opportunistic cooperation against it.

International causes of failure and success

Historically, illiberal paths to modernity have led to dead ends in part because of limitations of the institutions they use to stimulate economic growth. The Soviet collapse was mainly from these sources.

Other illiberal modernizers, especially Germany and Japan, failed largely because of the geopolitical costs of a system of political authority based on unchecked state power, the justification of rule by nationalist ideology, and a foreign policy aimed at the direct control of resources and markets through military conquest. These regimes provoked their demise not simply because they were expansionist, which is all too common in international affairs, but because they were heedlessly expansionist, gratuitously provoking overwhelming opposition and failing to learn to retrench from their overcommitments.

These mistakes were not incidental to their institutions and ideologies of illiberal late development. The seminal economic historian Alexander Gerschenkron (1943) showed how late, copycat development requires centralized financing and policy leadership by a strong state, which leads in turn to authoritarian political coalitions and the cooptation of mass support through nationalist ideology.

Admittedly, the errors and recklessness of Germany and Japan were stimulated in part by the shortcomings of the liberal international system in which they were operating. By the mid-1920s, both of these late-developing great powers were governed by labor-export coalitions in fragile, somewhat democratic multiparty systems, cooperating internationally in security and economic arrangements with the major liberal states. This was the era of the Locarno security treaty in Europe, the Washington naval arms limitation treaties, and the Dawes and Young plans for financing war reparations. However, the collapse of international financial stability and free trade in the Great Depression undercut the cooperative diplomacy of these coalitions, which were replaced by popular authoritarian nationalist regimes that sought direct political control over resources and markets in expanded empires. Middle classes and industrial cartels turned out to be fickle liberals, supporting free trade in consumer goods before the crash and shifting into imperial strategies requiring military production after it (Snyder 1991: 112–20, 133–50). In late developers, a weak capitalist class and middle class is often tempted to enter into an alliance with the state and traditional elites, in part to get protection from a restive and growing working class. In Germany, a shift of the capitalist class from support of free trade to protectionism happened twice, at the beginning of the 1870s and the end of the 1920s, both times triggered by sharp downturns in international markets (Gourevitch 1986: 71–180).

More broadly, an important predictor of a peaceful, successful transition to democracy is having stable democratic neighbors: be Spain, not Burundi. More generally, in recent decades the power and number of democratic states in the international system has become increasingly important in determining the likelihood of a successful transition, compared to the greater importance of the domestic characteristics of the state in earlier eras (Haggard & Kaufman 2016: 129–30; Gunitsky 2014).

In short, the strength of the liberal order and the ease of rising powers' transition to modernity are interdependent. This leads to a conundrum. If the "G-Zero" school of thought is correct, middle-income rising powers will act opportunistically at the expense of the prevailing liberal economic and security order. The resulting weakening of global liberal institutions could in turn weaken the economic and political position of those social interests in middle-income great powers that depend on the smooth functioning of the liberal system, intensifying the syndrome of authoritarian populist nationalism. Cooperation among the developed democracies in keeping global regulatory institutions strong is a top priority for managing the rise of the illiberal powers.

4

VARIEGATED CAPITALISM AND VARIETIES OF MODERNITY

Tak-Wing Ngo

Introduction

Debates about modernity have fallen into a circular trap. An earlier optimism that saw modernization as a homogenizing process leading to the convergence of societies and the universalization of the Western liberal order is now considered at best naive, if not ideologically hegemonic. Advocates of multiple modernities identify different pathways to, and distinctive forms of, modernity as a result of diverse historical trajectories and cultural backgrounds. Thus began the controversies over whether there is one modernity or many, and whether there are multiple modernities or varieties of modernity. The main query concerns whether there are modernities that cannot be understood fully in terms of the categories and concepts developed from Western experiences. As Eisenstadt (2000: 3) argues, modernity should not be equated with Westernization. Other forms of modernities exist, such as those in China and Japan. The latter represent not merely variations of Western modernity, but autochthonous and autonomous types of society. Ironically, even if we find this position taken by Eisenstadt and other advocates of multiple modernities to be strong and "politically correct," those alternative modernities have never fully conceptualized. If there are indeed multiple modernities, which societies are modern and which are not?

Varieties of modernity or multiple modernities?

The problem is that there is little consensus on the defining conditions of a modern society. Critics have rightly observed that we tend to take the development of Europe and the United States as prototypical of modern society. Even Eisenstadt (2000) has to admit that Western modernity enjoys historical precedence and is therefore taken as a standard. As such, secularization, scientism, industrialization, democracy, and a liberal market economy are often regarded as the major points of reference. This Eurocentric view increasingly has faced challenges, even among believers in a single modernity. From the outset, enormous diversity exists within Western nations in the way their societies, economies, and polities are organized. In the meantime, even when the economic, political, and ideational institutions originating in the West have spread across the world as desirable ideals and norms, their reproduction, adaptation, and modification are shaped by indigenous social, cultural, and cosmological practices under different civilizational traditions. The question is: Should these hybrid forms deriving from diffusion and adaptation ever be considered modern? If we drop the Eurocentric view, on what basis can we define modernity, given the implicit assumption that the West represents a clear type of modernity? What kinds of measuring rods can we find besides Western experiences that would allow us to construct a non-Eurocentric conception of modernity?

Some scholars have tried to escape this trap by referring to abstract conceptions of modernity. For instance, Wagner (2001) highlights human agency in realizing self-autonomy and mastery in setting and understanding the rules of human society as the condition of modernity. Wittrock (2000) argues that modernity is a set of promissory notes of hopes and expectations. Unfortunately, whether conceptualizing modernity in terms of human agency and promissory hopes can actually avoid Eurocentrism is debatable. From a skeptical frame of mind, this strategy merely retreats from using Western institutions as the reference point to highlighting Renaissance values as the condition of modernity. As observers have noticed, human agency based on individualization is a historically and culturally special form of modern development characteristic of the European experience rather than representing a universal logic (Yan 2010).

Beck and Grande (2010) rightly criticize Eisenstadt's formulation of multiple modernities for an insufficient elaboration on their structural variations and for a failure to see that various modernities are interwoven into complex relations of dependence from the outset, rather than being the products of internal evolution under a closed system. Beck and his colleagues have advocated the idea of "methodological cosmopolitanism" to overcome the Eurocentric account of modernity (Beck 2006; Beck & Grande 2007). They view methodological cosmopolitanism as "an approach which takes the varieties of modernity and their global interdependencies as a starting point for theoretical reflection and empirical research" (Beck & Grande 2010: 412). Thus, they seem to adhere to a temporal conception of modernity, as they refer to our time as a Second Modernity when the politics of "world risk society" has rendered national states and international organizations obsolete. In their words, "we all live in a Second, Cosmopolitan Modernity – regardless of whether we have experienced First Modernity or not" (Beck & Grande 2010: 418). As such, cosmopolitan modernization underlines not only varieties of modern society but also the mutual interaction and hybridity among societies.

This temporal conception of modernity may bring us back full circle. Wittrock (2000) distinguishes a temporal conception of modernity from a substantive one.

By "temporal conception" he refers to the modern age as a distinct epoch in world history. In this sense we are all living in an encompassing historical age. Although there are variations in the institutions and practices across nations, these are at best varieties of modernity, rather than fundamentally different forms of modernities. However, to allow us to speak of a modern age in the first place, we need to delineate the specific cultural and institutional features that are shared by modern nations. Hence the attempt to formulate a "substantive conception," which identifies certain key defining institutions and behaviors as modern. In this inquiry, whether a nation is modern or not, or whether certain orders within a nation are modern or not, is essentially an empirical question.

In retrospect, the absence of empirically-informed theorization lies at the core of the circular debate. So far, the debate has been fueled by abstract theorizations rather than observable comparisons of concrete cases. Few writings about multiple modernities have detailed precisely what sets one modernity apart from the other. As Schmidt (2006) skeptically asks, what justifies the language of "multiple modernities" rather than "varieties of modernity"? If we take the idea of methodological cosmopolitanism as our starting point for investigating the empirical varieties of modernity and their global interdependencies, can we forcefully present a coherent case of some alternative modernity that embodies similar degrees of hope and expectation and human agency as the Western model?

Capitalism and modernity

In parallel to the controversies about multiple modernities, scholars have been debating the varieties of capitalist order that characterize a modern economy. It is now widely recognized that, at a given historical period, various forms of capitalism may coexist. What exactly are these forms of capitalism? How do they differ? Are they variations or variegations of orthodox capitalism?

The debate over the varieties of capitalism has major implications for our understanding of modernity. This is because capitalism is closely related to modernity. As Sayer (1991) puts it, capitalism is the breeding ground for distinctive forms of modern sociality and new kinds of individual subjectivity. I have no intention of equating capitalism with modernity, as some observers have gone a long way to reject the conflation between the two (Wood 1997). But capitalism is undeniably the most universalizing trend – or, in Weber's words (1930: 17), "the most fateful force in our modern life." Critics have underlined the world-wide and totalizing character of the capitalist form. If a capitalist form of order exists in variegations, we can take that as the point of departure to establish empirically-possible varieties of modernity.

Despite the proliferation of debate and discussion on varieties of capitalism, the core of the debate remains firmly focused on the experiences of Western countries. Most of Asia, with the exception of Japan, has been ignored in this debate. The exclusion of the Asian experience once again reveals the Eurocentrism in existing scholarship. Bringing the Asian experience back into the debate will be the first step towards cosmopolitanism. The rise of Asia gives us a good opportunity to examine the possible varieties of modernity. I will use the case of China to illustrate this point. I argue that China offers new understandings of possible ways of organizing economic life that go beyond the narrow conceptualization of capitalism in existing scholarship. The Chinese case opens new windows to see the world from an alternative perspective.

Before we discuss the case of China, let us first take a brief look at the varieties of capitalism debate. Rejecting a monolithic conception of capitalism, scholars who study this issue believe that different forms of capitalism, including the American model of competitive capitalism, can coexist in the long run. They set out to theorize varieties of capitalism, and contrast the institutional similarities and differences between capitalist economies. The outcome is a dichotomized distinction between liberal and non-liberal capitalism (Streek & Yamamura 2001); or between the neo-American and Rhine models (Albert 1993); or between the liberal market economy and the coordinated market economy (Hall & Soskice 2001b). Despite differences in terminology, these studies mainly juxtapose the various ways in which economic exchanges are coordinated in the United States and continental Europe. The most influential work is the collection edited by Hall and Soskice (2001a). They claim to provide "a new framework for understanding the institutional similarities and differences among the developed economies," and to analyze the strategic behavior of economic actors under specific institutional and regulatory environments. Subsequent proponents of varieties of capitalism have also focused their attention on the firm, and consider companies to be the "crucial actors in a capitalist economy" (Hall & Soskice 2001b: 6). Hall and Soskice seek to show that the liberal market as characterized by the American model is neither the only nor the most efficient configuration of capitalism. They contrast it with the German model which they theorize as market-coordinated capitalism.

Over the years we can observe increasing nuance and sophistication in the conceptual frameworks and typological exercises developed within the varieties of capitalism debate. Hollingsworth and Boyer (1997) reject the long-standing myth of the market as the sole institution of economic coordination. They identify a myriad of other modes of coordination, including the company, the community, the state, the network, and the association. They focus on the ways capitalist economies combine market and non-market modes of coordination to sustain geographically and historically distinctive systems. Combining a conception of the different modes of coordination with a framework used in regulation theory, Boyer (2005) argues that there are at least four forms of capitalism coexisting in OECD countries. These four capitalisms are:

- A market-oriented capitalism typically found in English-speaking countries.
- A meso-corporatist capitalism, driven by the exchange of solidarity for mobility, as in the case of Japan and Korea.
- A state-driven capitalism, where innovation, production, and so on are shaped by public interventions, as is the case in continental European countries.

- A social-democratic capitalism, based on frequent negotiations between social partners and public authorities, as found in the Scandinavian countries.

Such refined categorization is a welcome addition to the existing debate. However, it is at best involutionary theorization, since it keeps adding analytical richness and complexity, and increasing theoretical sophistication, at the risk of ignoring and excluding the larger and more diverse set of experiences in parts of the world that seem to go beyond the conceptual grasp of the West. As Peck and Theodore (2007: 761) put it: "Where should we locate China, India, and Brazil in this picture? And what do their modes of growth reveal, not only about the bandwidth of contemporary capitalist 'variety,' but also about the interpenetrating nature of capitalist development at the global scale?"

Post-socialist neo-capitalism in China

In this regard let us examine the case of China. To do so let us first ask the very basic question: To what extent can we consider China a capitalist society? Pontusson (2005: 164) is correct to point out that the literature on varieties of capitalism has a great deal to say about varieties, but surprisingly little to say about capitalism. If we follow the suggestions of some of the classical works, we find that, despite the official claim of a socialism with Chinese characteristics, contemporary China falls into the capitalist category. For Marx, capitalism was defined by the commodification of goods, services, and labor. Capital accumulation is generated by the production and exchange of commodities, which becomes the engine of growth of the economy. Echoing this view, Polanyi (1944) argues that capitalism is driven to extend the logic of the marketplace throughout the whole of society, such that social relations are monetized and organized in terms of market relations. The "great transformation" occurred not simply in the creation of capitalist institutions and their related political and legal frameworks, but more importantly in the fundamental alteration of human mentalities. Eventually, anything can become a commodity under capitalism, from a dating service to Shaolin martial arts – a situation that Marx labeled "fetishism." While Marx believed that the single-minded drive for profit would lead to over-production and the eventual demise of capitalism, Schumpeter (1934) regarded entrepreneurial innovations as the feature that allows the capitalist system to rejuvenate itself despite periodic downturns. The problem of over-accumulation in current times, according to Harvey (2004), finds its solution in accumulation by dispossession. Accumulation by dispossession can occur in a variety of ways; its modus operandi is thus contingent and haphazard. It can include the appropriation of communal lands and the expulsion of the peasantry, the conversion of communal property rights into exclusive private ownership, the replacement of indigenous forms of production, asset-stripping through mergers and acquisitions, raiding of pension funds, corporate fraud and stock manipulation, and so on. These numerous forms of dispossession are everyday occurrences in China. Eventually, according to Harvey, if surplus capital and labor cannot be absorbed domestically, then they will be sent elsewhere to find new terrain of profit realization, as represented by China's vast investments in Africa and the grand strategy of "One Belt One Road" (aka the Belt and Road Initiative).

In the wake of extreme commodification, a single-minded drive for accumulation, constant innovation, and ruthless speculation – not to mention the active part it plays in global market exchanges and the international division of labor – it is difficult to perceive China as anything other than a capitalist economy. But if China is a capitalist economy, it exemplifies a kind of capitalism that is not familiar to the West. More fundamentally, its operating principles do not follow the textbook model of the capitalist market economy. Nor can it be placed comfortably in any one of the categories within the varieties of capitalism typology. I therefore call it post-socialist neo-capitalism, to distinguish it from the orthodox form of capitalism including its variations.

China's capitalist system distinguishes itself from the Western counterparts in at least five fundamental areas: its economic players, ownership structure, state–market relations, rules of exchange, and nature of competition. These are the constitutive dimensions of an economic order. The idiosyncrasy of China's neo-capitalism in these five dimensions renders the differences between the liberal and coordinated-market variants completely trivial. I will outline each of these dimensions briefly.

1. Economic players

Literature on varieties of capitalism urges us to look at three major actors: the government, firms, and workers. The firm is seen as the main embodiment of capital. In China, companies are not the sole capitalist actors in the economy. A large variety of state actors and social actors, as well as individuals, stand side by side with enterprises in business operations. State actors include not only the government bureaucracy but also the ruling party, the military, state-sponsored people's organizations, and so forth. All have business connections. Social actors such as universities, news agencies, and hospitals have business ties. It is an exemplary case of extreme commercialization and commodification, to the extent that religious bodies, rural villages, educational institutions, medical establishments, etc. are run as profit-making enterprises. Furthermore, a very substantial proportion of economic activities is taken up by individuals or networks of individuals in underground finance, production and distribution of counterfeit products, trafficking, cross-border trade and exchanges, sex services, and so on. They are not necessarily illegal, as huge volumes of parallel trade, Internet shopping services, and the like form a close symbiotic relationship with legitimate economic activities in the provision of daily products and services (Hung & Ngo 2015). Common sense tells us that to a certain extent these actors and activities can be found in the West as well. Unfortunately, the dominant paradigm of neoliberal political economy has chosen to exclude them completely from analysis, as if they did not exist or merely played a negligible role. What remains in the analysis is the undisputed role of the firm as the only decisive actor. The firm is seen as the triumphant player that minimizes transaction costs, maximizes competitive competencies, and develops innovative capacities. However, as Braudel (1982: 432–33) reminds us, an essential feature of the general history of capitalism is its unlimited flexibility, its capacity for change and adaptation. And the source of that extreme flexibility comes from the real capitalists who refrain from specialization and professionalization, and hence avoid being bound by restrictive rules, practices, skills, and networks in a particular sector. For Braudel, it is not just Schumpeterian entrepreneurship that matters; it is cunning, intelligence, and, above all, power that matter. China's remarkable growth rate in recent history is in no small part attributable to state-cum-capitalist actors and social-cum-capitalist actors freely switching their functions between public service and profit-making, and manipulating the system in their entrepreneurial activities.

2. Ownership structure

Even if we put our focus exclusively on the firm, the case of China still defies the understanding of the existing paradigm. This leads us to the second area of departure: ownership structure. In neoliberal political economies, there is a clear distinction between the public and private sectors. The market is dominated by private firms with a well-defined ownership structure engaging in mutual exchanges for the sake of profit maximization. That is why it is often referred to as the private sector. In China, property rights are highly diffused, with different degrees of overlapping public and private ownership. This is not the same as a mixed economy where one can find both publicly-owned enterprises and private firms with well-defined boundaries. Instead, there exist in China a vast number of companies which cannot be clearly categorized as either public or private.

From the outset, administrative restructuring has led to the commercialization of a substantial number of bureaucratic units into companies or professional social organizations, in addition to the vast number of state-owned enterprises. Many of these quasi-state entities served as intermediaries between enterprises and state agencies. They were neither fully public nor entirely private. They sought profit but simultaneously took political orders from their supervising state units. The proliferation of quasi-state commercial establishments blurred the public/private divide. State agents frequently became business managers. This blurring has given birth to a variety of actors, a diversity of organizational forms, and a plurality of property rights. In this unique constellation, the most prominent actors are not conventional private businesses, but a new breed of bureaucratic enterprises, politically embedded businesses, joint public–private ventures, and private firms disguised as collective enterprises. Various forms of ownership exist side by side. Property rights have been divided into components such as right of usage, claim to resource flow, and right of transfer, with different ownership structures commanding different combinations (Ngo 2011).

There is a dearth of conceptual categories to describe the plurality of property right regimes, since the dominant paradigm finds little need to draw such distinctions. However, the Chinese case is not the only exception. The blurring of the public/private divide and the fuzziness of enterprise boundaries are defining features of Eastern European capitalism as well. Stark (1996: 1016) calls this "recombinant property," where new forms of ownership emerged when the qualities of private and public are dissolved, interwoven, and recombined, and firms have exploited such ambiguity to their advantage. It reminds us again of the unlimited flexibility that Braudel described.

It is important to understand the nature of recombinant ownership, because different ownership arrangements lead directly to different interests and forms of rationality. Many of these businesses respond simultaneously to market signals and political incentives. For instance, enterprises with substantial state ownership are often entrusted with multiple tasks in addition to profit-making. Some are pressured by local governments to invest excessively in order to achieve higher GDP growth. Some are asked to take on more workers, regardless of their actual production capacity, when unemployment rises. Some profitable enterprises are obliged to take over or merge with money-losing ones (Buckley et al. 2005).

3. State–market relations

This brings us to the third area of deviation: state–market relations. It is widely recognized that the Chinese state is not only a policymaker and referee. Similar to the experience of other East Asian countries, the state in China plays an active role in steering economic development and in "leading the market." More than that, the Chinese state is also an investor, an active market player, and a profit-making entity. The regulator-cum-player role of the state inevitably shapes the operation of the market. The market is anything but a level playing-field where transactions are conducted between equal parties. And this is just the tip of the peculiar state–market relations in China. During the reform period, China has devolved much of the regulatory, fiscal, and administrative power to local governments at various levels and to regions. The result is a decentralized and fragmented system with the locus of agency resting not only on the central authorities but also on the sub-states in various provinces, prefectures, counties, and townships. These sub-states compete fiercely for resources and investments, and deploy various measures to protect their local markets vis-à-vis other regions. This in turn leads to the fragmentation of the market. Under the logic of local protectionism, sub-states are often prepared to promote local development at the expense of the national economy. In other words, capitalist development in China is very even as a result of specific state–market relations. Cataloguing the system at the national scale is at best a gross misrepresentation of its institutional rationality (Ngo et al. 2016). The "dynamic polymorphism" of Chinese capitalism, to borrow Peck and Theodore's term (2007: 761), underlines the scalar specificity and spatiality of capitalist development – important dimensions that have been ignored in the varieties of capitalism debate.

4. Rules of exchange

The fourth area of dissonance concerns the rules of exchange. Many economic transactions in China do not fall under the category of "free" contractual exchanges between firms. Instead, market players, including firms with varied ownership structures, have crafted their conventions of socio-economic exchange rooted in shifting state power, cultural practice, and personal ties. As a result, a substantial proportion of business transactions is based on exchanges at the personal rather than enterprise or organizational level. This kind of exchange relies on a cultural understanding of trust, obligation, and reciprocity, which is established through the mutual exchange of gifts, favors, banquets, and other symbolic artifacts. In this realm of exchange, there is little distinction between material transactions, symbolic offerings in the form of face-giving, and the discharging of social obligations in honor of friendship, kinship, or clanship (Yang 1989). The cultivation of personal networks (*guanxi*) has become a common strategy for companies, domestic and foreign alike, doing business in China.

Likewise, a substantial proportion of economic resources is allocated not by the mechanism of supply and demand, but by rent-seeking. Given the extensive degree of state intervention and the shifting roles of the state as regulator-cum-investor, an enormous amount of economic rent is created and transacted. As a matter of fact, the creation and allocation of economic rents have been used by the Chinese state as a policy instrument in effecting industrial plans and development priorities. By controlling resources such as low-interest loans, foreign exchange, and the like, and by exercising its power to ration these resources to targeted industries or enterprises, the state, in both its central and local institutions, has been the main creator of rents and particularistic benefits (Ngo 2008). The availability of enormous social resources as rents induces economic players to engage in rent-seeking. These rent-seekers are not confined to enterprises, but include the whole range of players such as state bureaus, local governments, public bodies, social organizations, and individuals. Elsewhere I have developed a typology of different types of rents and their terms of exchange (Ngo 2009). Suffice it to say here that this common form of transaction in China has its own distinctive rules. It is far from arbitrary, and cannot be equated with sheer corruption. Theorists of public choice used to see rent-seeking as nothing more than a pathological deviation from free market exchange. In China, it is a highly institutionalized transaction that cannot be eliminated from the research agenda by definitional fiat.

5. Nature of competition

Finally, we turn to the nature of competition. If capitalism is characterized by its strong capacity to rejuvenate itself, competition is the driving force of such innovative adaptability. So far, competition has been conceptualized largely in terms of competition among firms for profit. Under "free market" conditions, firms are believed to be under a constant pressure to increase production efficiency, reduce costs, and innovate in order to maintain their competitive edge. For that reason, monopoly has been condemned as a pathological obstruction to free competition – one of the exceptional situations in which state interventions are deemed appropriate to restore free market exchange.

China again prompts us to reconsider this conceptualization. On the one hand, intense competition can be found in the Chinese market. On the other, monopolies and oligopolies abound. Even more intriguing is that cut-throat competition often exists between oligopolies or even within a monopoly (Cheng & Ngo 2014). Moreover, the goal of competition is not reducible to profit maximization. The latter is only one of the main objectives of firms in China. Other objectives of competition include increases in market share, scale of production, performance targets, and ranking. Apparently, these different goals of competition have provided impetus for innovation and adaptability. However, the implications for the market structure, growth pattern, labor relations, and so forth are vastly different from the kind of competition that focuses exclusively on profit maximization.

Conclusion

In their seminal work on varieties of capitalism, Hall and Soskice (2001b) underlined a number of dimensions of difference for sorting out the dichotomized models of capitalism. These dimensions include corporate structure, financial system, education and training regime, industrial relations, and inter-firm relations. We have seen that Chinese capitalism far exceeds and exhausts the reach of such classification. Differences in terms of the nature of economic players, ownership structure, state–market relations, rules of exchange, and nature of competition are at least as crucial as, if not more fundamental than, those dimensions raised by Hall and Soskice in underlining capitalism as a mode of exchange and production.

The idiosyncrasy of Chinese capitalism allows us to expand the conception of economic systems beyond the socialism/capitalism dichotomy as well as beyond categories of varieties of capitalism. The comparison between the Rhine model and the American model simply neglects larger differences within the capitalist family. Zhang and Peck (2016: 55) make this point elegantly:

> Cases like China, which do not seem to fit the black-and-white distinctions between the LME and CME ideal types, are likely either to be misconstrued or to be sidelined as *sui generis* exceptions if the varieties sensibility is deployed unreflexively. It is possible that not even shades of grey will suffice; some cases should prompt a revisualization of the spectrum itself.

Of course, one can argue that Chinese capitalism is at best transitional, and cannot therefore be treated as a mature prototype. I fully agree that the economic system in China is transitional, but it is no more transitional than welfare capitalism in Europe or Fordist capitalism in the United States. As Braudel insisted, unlimited flexibility and adaptation are the essence of capitalism. Echoing this point, Boyer (2011: 70) reminds us that it is erroneous to look for static properties of capitalism, since it is by nature a constantly evolving regime. One thing, however, is certain. Even if the Chinese system is transitional, its direction of change is unlikely to move towards the Western varieties in the foreseeable future.

If China presents a critical challenge to existing categories and conceptions, the addition of experiences from other Asian countries will open even more windows for alternative theorization. Added together we may question the universal logic of Western capitalism, and show that it is at best a partial representation of the rich varieties of modern economic order.

The same exercise can be taken to examine the social, political, and ideational orders that point to alternative modernity projects. In fact, some studies have already embarked in such endeavors. For instance, Yan (2010) has documented the unique path to individualization in China. With the accumulation of efforts, we may be able to piece together coherent cases of alternative modernity beyond the Western paradigm.

PART III

Human security

5

MULTIPLE MODERNITIES IN A MULTIPLEX WORLD

Amitav Acharya

A global idea-shift

The election of Donald Trump as the President of the United States of America raises question marks over the future of the liberal international order. To be sure, uncertainty about the fate of the liberal order had been anticipated for some time (Acharya 2014). But until now, it was generally assumed that the main challenge to liberal order would come from external factors, viz. the rising powers led by China. But Trump's victory, especially in the wake of the Brexit decision, suggests that the main challenge to the liberal order comes from within (Acharya 2017a).

Although the Trump presidency, along with Brexit, poses an internal challenge to the liberal order, the latter's relative decline cannot be delinked from the global power shift. A few trends here are revealing. In the year 1800, China alone produced about 30% of the world's Gross Domestic Product. Add India, and that figure rises to more than half of the world's GDP back then. And strikingly, the age of Asian economic primacy may be returning. A recent report by the Asian Development Bank suggests that by 2050, China alone could account for 20% of the global GDP and India for 16%. By comparison, the US will account for only 12%. There is also evidence of a more general "rise of the rest." The non-Western countries (minus Japan) have increased their share of global output from one-third in 1990, to about half today. The Global South's share of the global GDP rose from 33% in 1980 to 45% in 2010 (UNDP 2013: 2) while extreme poverty (below $1.25 per day) went down from 43.1% in 1990 to 22.4% in 2008 (UNDP 2013: 13). The OECD estimates that the Global South could account for 57% of global GDP by 2060 (*The Guardian* 2012; KPMG International 2014: 34). While growth in some of the emerging powers has slowed down in recent years, much of the shift has already occurred and will continue to redefine the global environment. "The global economic power shift away from the established, advanced economies in North America,

Western Europe and Japan will continue over the next 35 years" (Price Waterhouse Cooper 2015: 1). This could mean a future in which globalization is driven more by South–South linkages than by North–North or North–South linkages (Acharya 2017a).

The decline of the Western-dominated world order, which I have called the American world order, is not leading to multipolarity, as many traditional pundits assume, but to *multiplexity* (Acharya 2016a; Kuo 2016). Multiplexity, or the idea of a multiplex world, has the following main features. First, it is not a multipolar world in the traditional European sense. That order assumed the primacy of the great powers, whereas a multiplex world will have many different actors. These actors are not just states; rather, they also include international institutions, non-governmental organizations, multinational corporations, and transnational networks (good and bad). They are connected by complex forms of linkages that include not just trade but also finance and transnational production networks, which were scarce in pre-World War European economic interdependence. Moreover, interdependence today goes beyond economics and also covers many other issue areas, such as the environment, disease, human rights, and social media. A Multiplex World has multiple layers of governance, including global, interregional, regional, domestic, and sub-state.

Last but not the least, as with a multiplex cinema, a multiplex world gives its audiences a wider choice of plots or stories. Some of these plots – or ideas and ideologies – differ from and challenge the cultural and political narratives and instruments of liberal modernity, and its offshoot, the US-dominated liberal international order. There is no "end of history" here, except in terms of the relatively short history of Western dominance.

Hence another challenge to prevailing international order comes from the realm of ideas. The global power-shift has been accompanied to some extent by a global idea-shift (cf. Acharya 2016b). In a Multiplex World, the West no longer dominates the global idea "marketplace."

One of the most problematic features of liberal modernity, and the liberal international order that it underpins, is its underlying assumption that the West is the only or main source of progressive, modern ideas. Western ideas are assumed to have a timeless quality. These include ideas such as nationalism, democracy, human rights, capitalism, development, and even the more controversial idea of national security. They are held to be universal standards that apply to all of humanity and should never be compromised or challenged. The role of the non-Western world is one of passive acceptance. Whenever there is a challenge to such Western ideas, it is dismissed as radicalism or foolhardiness.

Furthermore, proponents of liberal modernity assume, whether explicitly or implicitly that Western ideas should and do dominate and displace the ideas of other parts of the globe, especially the bad ideas and practices of the Rest. If good ideas are occasionally found outside the West, they are dismissed as imitations. If they prove to be creditable, the credit is usually given to the Western training of the person who invented the idea or to Western governments and institutions that backed them. One way or the other, somehow a Western origin or connection is found to legitimize ideas from non-Western sources.

Many people still think the West will continue to be the main springboard of new ideas about governance, development, and peace. But a quiet revolution is taking place that challenges this conventional assumption. The non-Western world increasingly is coming up with critical ideas about governance, development, and security. It is not simply a passive recipient of Western ideas, but an active contributor to the world's stock of ideas on these matters.

This global idea-shift has several features. First, unlike in the past, when most ideas from the non-Western world were about resistance or rejection (e.g., against colonialism, neo-colonialism, or superpower intervention), the new ideas concern problem-solving, especially in regard to the common problems of humankind. For example, the anti-colonial struggles produced a whole slew of ideas from the colonies. We originated the Non-aligned Movement (NAM) and the New International Economic Order (NIEO). These ideas made a huge splash on the world stage, but also caused a sharp North–South divide. By contrast, the new ideas come with a much more positive message and tone. They are aimed at bridging the North–South divide.

Second, they are not necessarily coming from the rising powers such as China, India, Brazil, etc., or the rising regions like East Asia, but also from far less fortunate places like Bangladesh, Pakistan, and Africa.

Third, new ideas are emerging because the big ideas of the past have failed us. While ideas such as democracy, human rights, capitalism, development, and peace are held to be universal standards that apply to all and should never be compromised or challenged, these big ideas sometimes still have big holes. They promise one-size-fits-all solutions. Do they fit all parts of the world? The ideas of human rights and democracy may seem genuinely universal, but does it follow that they are the answer to the problems of governance in all parts of the world? No one would object to these ideas, or ideas of development, security, and democracy. But these big pictures and principles often offer a poor fit for many parts of the non-Western world. To some, this may sound like cultural relativism. In reality, it is not about culture, but about practicality.

In a recent book, Steve Weber and Bruce Jentleson (2010) argue that the core ideas of the West, ideas deeply associated with liberal modernity, including capitalism and democracy, are neither universal nor durable. The authors may be going too far in assuming that the ideas of peace, democracy, and development that define contemporary liberal modernity might be displaced. *What might disappear is the West's monopoly over how peace, democracy, and development are interpreted and pursued.* That eventuality would bring out the multiple, global heritage of these ideas, thereby recognizing and encouraging a rich diversity of human understanding and action that can only be good for our world.

To sum up, the global idea-shift features key ideas about development, security, and ecology coming from the Global South. Some of the biggest innovations for helping free humanity from the scourge of poverty, injustice, and conflict have come from outside of the West, from places where they are most needed. Because their pioneers are often trained in Western academic institutions and their innovations disseminated through Western think-tanks and Western-dominated global institutions, they are often hidden from the public eye. Sometimes the credit for these innovations goes to the Western governments and international agencies that market these ideas. In reality, the West often plays a skeptical, or at best supportive, role. The new *ideapolitics,* or the spread of ideas originating in the non-Western world and addressing the conditions and concerns of the vast majority of humankind, may well shape the 21st-century world order.

The idea of human security, which has been around for over a quarter-century, offers a striking example of the global idea-shift and a powerful challenge to the core assumptions of liberal modernity. This essay discusses the origins and implications of the human security idea. It argues that although the human security idea is post-Western and post-liberal, reflecting the strong local (South Asian) context of its creators, it has won global recognition and application. Moreover, the idea challenges two of the core ideas of liberal modernity: its association of economic development with growth of the Gross Domestic Product (GDP) and its conflation of security with "national security." It thus offers a powerful example of the notion of "multiple modernities."

The human security challenge

There have been major disagreements over how to define human security. According to the UNDP, "Human security can be said to have two main aspects. It means, first, safety from such chronic threats as hunger, disease and repression. And second, it means protection from sudden and hurtful disruptions in the patterns of daily life – whether in homes, in jobs or in communities." For the Canadian government, which has played a key role in promoting human security, "human security means freedom from pervasive threats to people's rights, safety or lives." The Government of Japan defined human security "as the preservation and protection of the life and dignity of individual human beings." And the UN Commission on Human Security, saw the objective of human security as to safeguard the "vital core of all human lives in ways that enhance human freedoms and human fulfilment" (all cited in Acharya 2017c).

Early on, a major tension emerged between two conceptions of human security: freedom from fear and freedom from want. The Canadian formulation saw human security as a matter of reducing the human costs of violent conflict and sought to pursue it through policies geared to abolish land mines (the Ottawa Land Mine Treaty), deter and punish genocide and crimes against humanity (through the creation of the International Criminal Court), and curb the recruitment of child soldiers. It rejected the developmental aspects of human security, such as those related to poverty reduction. It was concerned that including social and economic measures of human security would make the concept too broad, thus implying commitments that the scarce resources of the promoters of human security would not be able to meet. But the Japanese and Thai (then under democratic rule) conceptions of human security insisted on including economic and social aspects within the security paradigm. After a period of debate, there is now overall consensus that human security should include both freedom from fear and freedom from want. As I stated in a speech to the UN General Assembly (Acharya 2011):

> The evolution of the concept of human security has gone through two phases. Between the 1990s and the early 2000s was a period of debate over its various meanings, whether human security was about freedom from fear, or freedom from want. Since then, the debate has entered a period of general agreement that human security is both, as well as freedom to live a life of dignity. It is not a matter of either this or that. The important challenge is how to look for linkages between these various meanings.

The initial contestations over the definition of human security tell a story that relates to my earlier point about the tendency of the proponents of liberal modernity to claim a monopoly over the genesis of good, progressive, and problem-solving ideas. The fact that the government of Canada, as well as its partner in the effort, Norway, claimed leadership in the human security movement and sought to promote a narrow conception of human security was neither new nor exceptional. Yet, this was without regard for the fact that the idea of human security originated in the Global South. It was not a *Western* idea. Human security was an outgrowth of the human development idea, whose original articulation came from a Pakistani development economist, Mahbub ul Haq, who was once the Finance Minister of the country. It was supported by the ideas of Amartya Sen, who was born in a part of India that now belongs to Bangladesh.

The human development idea rejected the deep association between development and GDP growth, a key element of the international liberal order. Haq argued that GDP growth alone will not satisfy the basic needs of the world's poor. True development depended "on the quality and distribution of economic growth, not only on the quantity of such growth" (Haq 1995: 15). High economic growth does not always trickle down to the people; thus, more efficient policies should be implemented to expand the well-being of a country's citizens. Income is important, but should not to be equated with human well-being. Also important are political freedom, the environment, health, and education. Hence, in considering development, instead of asking "How much is a nation producing?" the question should be: "How are its people faring?" And "the real objective of development is to enlarge people's choices" (Haq & Ponzio 2008: 8).

Amartya Sen's contribution to the human development and human security ideas was in distinguishing income poverty from capability poverty. He argued that while the orthodox idea of development stressed reducing income poverty, the real challenge to development lay in reducing capability poverty. Hence increasing a country's income (i.e., its GDP) does not necessarily increase the capabilities of its people. But without capability, there is no true, long-term development. A more effective approach to development, Sen proposed, would be to reverse the priorities by focusing on capability improvement first and going from there to increasing earning power (Sen 1999: 90). The idea of human development thus stresses expanding people's capabilities. The purpose of development is to enlarge all human choices, not just income. The human development paradigm is concerned with both building capabilities through investment in people and the full actualization of these capabilities (Momin, n.d.).

While Haq and Sen both received advanced degrees in the West, they drew from the context of their home countries and the region of South Asia. Haq was born in undivided India, while Sen was born in East Pakistan (later Bangladesh). The two were classmates at Cambridge University. Haq and his family fled the Indian part of Punjab in 1947, coming face to face with death in the communal violence that took nearly a million lives. Haq studied at Punjab University in Lahore for a degree in economics and then went to Cambridge. He eventually left Pakistan to work for the World Bank, becoming the director in the Policy Planning Department of the World Bank from 1970 to 1982. After a stint as Planning and Finance Minister of Pakistan (1982–88), he joined the UNDP as a Special Adviser in 1989, where he prepared the first Human Development Report in 1990. The two men formed a life-long partnership. It was during their conversations in Cambridge that Haq urged Sen that measuring development on the basis of one indicator alone, the GDP, is a "silly" idea. While Haq went on to study for his doctorate at Yale, Sen got his PhD from Cambridge.

Haq's interest in a people-centric approach to development and security was inspired by his experience with India's partition. As Sen wrote: "the Pakistani experience of Mahbub influenced him. His family was from Kashmir; he nearly died at the time of partition" (Sen, n.d. a). Hence, "locality must have also played a part" in Haq's thinking (Sen, n.d. b). Although their message was intended for a global audience, South Asia formed a major context of Haq's and Sen's work. South Asia, Haq would write in 1995, was sinking "into a quagmire of human deprivation and despair." He was shocked that it had fallen behind sub-Saharan Africa, to become the most deprived region in the world. It was "tragically comic" that India and Pakistan, after "bleeding their economies" to pay for arms, "beg and submit to all sorts of conditionalities from international lending institutions." This sad state of affairs could be blamed particularly on the "guns versus butter" problem. High military expenditure came at the expense of a country's development. In his major work, *Reflections on Human Development*, Haq pointed out that the choices made by a country were an essential element of human development: "the national priorities chosen by society or its rulers – guns or butter, an elitist model of development or an egalitarian one, political authoritarianism or political democracy, a command economy or participatory development" (Haq 1995: 14) had an enormous impact on its future developmental path. Haq was also shaped by his experience in Pakistani government. While just after completing his Western education and serving as the chief economist of the Pakistani Planning Commission, he became aware of the fact that rapid economic growth did not translate into benefits for the poor.

On the contrary, the focus on growth might undermine those benefits. During 1982–88, his term in office serving as finance, planning, and commerce minister in Pakistan had the most "profound impact" on his thinking about approaches to development. His own policies to promote rapid growth had little impact on the conditions of the poor, prompting him to rethink his approach. The benefits of growth went to a few people. So, attention has to be given to human development.

It was in the 1994 "Human Development Report" issued by the United Nations Development Programme that the idea of human security was most clearly articulated for the first time (see Acharya 2001a). That report outlined seven main elements of human security:

- Economic security – an assured basic income for individuals, usually from productive and remunerative work, or, in the last resort, from some publicly financed safety net.
- Food security – ensuring that all people at all times have both physical and economic access to basic food.
- Health security – guaranteeing a minimum protection from diseases and unhealthy lifestyles.
- Environmental security – protecting people from the short- and long-term ravages of nature, man-made threats in nature, and deterioration of the natural environment.
- Personal security – protecting people from physical violence, whether from the state or external states, from violent individuals or sub-state factors, from domestic abuse, and from predatory adults.
- Community security – protecting people from the loss of traditional relationships and values, and from sectarian and ethnic violence.
- Political security – ensuring that people live in a society that honors their basic human rights and ensuring the freedom of individuals and groups from government attempts to exercise control over ideas and information.

From a people-centric notion of development came a people-centric notion of security. Just as the human development idea challenged the orthodox Western concept and model of economic development, the human security idea challenged the traditional Western notion of security. The latter was premised on the idea of national security, which has been and remains another core idea of the US-dominated, liberal international order. The idea of national security privileges the Westphalian nation-state, which is one of the key features of liberal modernity. Simply stated, national security means protecting the state, its sovereignty, and territorial integrity, from external military attack. The national security framework was developed in the West, especially in the United States after World War II. It reflected the quintessential American concern with the threat from the Soviet Union in the context of the Cold War. The national security approach was very Western-centric, in the sense that it paid little attention to the problems of developing countries, whose main threats come from domestic sources and socio-economic grievances and problems rather than from external military powers. Moreover, the main instrument of national security is military, or the use or threat of use of military force. The main instrument of human security is social, economic, and political.

In summary, the human security idea is distinguished by three core features: (1) its focus on the individual/people as the referent object of security; (2) its multi-dimensional nature; (3) its universal or global scope, applying to states and societies of the North as well as the South. A key innovation of the human security idea is its linking of development and security. It draws upon human development's stress on building human capabilities to confront and overcome poverty, illiteracy, diseases, discrimination, restrictions on political freedom, and the threat of violent conflict. Hence, "Individual freedoms and rights matter a great deal, but people are restricted in what they can do with that freedom if they are poor, ill, illiterate, discriminated against, threatened by violent conflict or denied a political voice...." (UNDP 2005: 18–19). As Haq himself suggested, "human development dealt with the betterment of human lives and ability for a community to thrive. Human security sought to highlight the levels of human development achieved, and to the 'security' of gains made by focusing on 'downside risks' of political conflict, war, economic fluctuations, natural disasters, extreme impoverishment, environmental pollution, ill health, illiteracy, and other social menaces" (Haq & Ponzio 2008: 114). Sen would call this the "promotion of security of daily life." Referring specifically to the impact of the Asian Financial Crisis of 1997, Sen (2000a) stated:

> Along with the old slogan of 'growth with equity' we also need a new commitment towards 'downturn with security,' given the fact that occasional downturns are common – possibly inescapable – in market economies. In achieving security under these circumstances, and in trying to guarantee secure daily living in general, we need social and economic provisions (for example, for so-called 'economic safety nets' and the guaranteeing of basic education and health care), but also political participation, especially by the weak and the vulnerable, since their voice is vitally important. This requires the establishment and efficient working of democracies with regular elections and the tolerance of opposition, but also the cultivation of a culture of open public discussion. Democratic participation can directly enhance security through supporting human dignity (more on this presently), but they also help in securing the continuation of daily lives (despite downturns) and even the security of survival (through the prevention of famines).

The human security idea's focus on the individual may seem paradoxical, since the West is traditionally associated with individualism and the non-West with communitarianism, relativism, and authoritarianism. But here is an idea from the non-Western world that puts the individual at the heart of both development and security.

Haq was initially given little credit for his innovation by Western policymakers. Then Canadian Foreign Minister Lloyd Axworthy, whose memoirs devote an entire section to Human Security, makes no mention of the origins of the concept in human development, or of its progenitor, Mahbub ul Haq. And, as noted, he tried to move the concept away from its economic roots in human development, and attempted to shift the focus to security aspects, such as banning land mines, the International Criminal Court, and humanitarian intervention. It is also ironic that while the non-Western advocates have championed human security in its broader meaning, many liberal states have downplayed, distorted, or narrowed the concept for strategic purposes. The US has virtually ignored the human security idea. National security remains paramount there. In many parts of the West, human security threats – poverty, inadequate health care, and environmental degradation – are not recognized for their own sake, but only when they result in violent conflicts that challenge national security. The US as the leader of the liberal international order offers limited protection to its poor against disease (no universal health care) and gun violence. And, as has been noted, there has been much instrumental "capture" or "misuse" of the human security idea by the Western liberal nations "in the guise of ethical concerns" to enhance their "state-oriented national interests," gain "greater influence in the United Nations, and increased credibility on the international stage" (Tadjbakhsh 2008: 131).

The end of liberal modernity?

The idea of human security thus offers a key challenge to the prevailing beliefs about liberal modernity. Liberalism, the ideological foundation of liberal modernity, is a slippery and self-serving concept. What is liberal and who is liberal is contested, even among the liberal scholars. More importantly, its past claims of openness and accommodation are exaggerated. During the Cold War, the liberal order was at best a limited international order, rather than genuinely global. It excluded key non-Western nations, such as China and India, as well as the Third World. It functioned mostly as a provider of "club goods" to America's allies and partners rather than as a supplier of universal public goods (Nye 2014: 1246–47). It expanded during the post-Cold War era, but can it earn the genuine loyalty of all? The legitimacy and the future of that order is not just a matter of the material benefits it confers on the rising powers, but also of ideational and identity considerations, including the resentment among the rising powers who associate the liberal order, its ideology, and institutions, with Western dominance and exploitation. As it stands now, the idea of a liberal international order coopting them evokes suspicion and even hostility from the emerging powers, including the democratic ones (cf. Nye 2017).

The liberal modernity idea is a broad concept that seeks to embrace all good things in life. Whether as an ideology or as a theory of world politics, liberalism is also inherently homogenizing. Despite its alleged respect for pluralism, it has not been particularly accommodative of diversity when it comes to locating ideas and agency in global order-building. (I would submit that realism and constructivism acknowledge more fully the diversity, autonomy, and agency of the non-Western states and societies.) Liberalism claims to be the source of all good things – human rights, free trade, institutions, and cooperation – yet its foundational narrative presents them as unique or distinctive contributions of the West, without much regard for the multiple and global heritage of these ideas.

The idea of human security draws attention to several limitations of liberal modernity. It mainly challenges the notion that all progressive ideas come from the West. While the idea of human development on which the human security idea is based was initially seen as a Western concept meant to improve conditions in the developing world, this is a misleading way to characterize it. In reality, the concept of human development challenged the core of liberal economics. Hence, it is not surprising that the venerable mouthpiece of Western liberal economics, *The Economist* magazine (n.d.) called Haq a "heretic" and his *Human Development Report* "a guidebook for economic heretics." The human development idea is people-centric: "Poor countries failed to prosper because they neglected the basic development of their people… [C]ontrary to the foundations of economic thought, people as the agents of change and beneficiaries of development were often forgotten by the development policy establishment" (Haq & Ponzio 2008: 8). And the idea of human development has been increasingly influential. The global ranking of nations based on this idea, the Human Development Index, was launched in 1990. Since that time, more than 600 regional and national human development reports (HDRs) have been published in more than 140 nations.

Human security also attests to two differing conceptions of universalism. The first may be described as "particularistic"; it is the essence of (liberal) universalism. It assumes, Enlightenment-like, that certain universal concepts such as capitalism, democracy, and human rights have a singular (European) origin, but should apply to all or have universal validity. The other notion may be termed "pluralistic universalism." Here, universality means having multiple and global origins. Human security falls in the latter category. It is South Asian in origin, challenging orthodox capitalism but adopted by the UN, thereby suggesting a different route to universality. As such, it is a prime exemplar of multiple modernities.

6

ENHANCED HUMAN SECURITY

A modernity to be available to all

Inge Kaul

Introduction

In recent years, issues of volatility, disaster, conflict, and violence have drawn added political attention, both in the nations of the North and the South and internationally. The greater prominence of these issues reflects how growth and development have been increasingly crisis-prone, with one crisis grabbing the political spotlight from the others. Judging from the recent *Global Risk Reports* of the World Economic Forum (2016, 2017), the world can ill afford another major challenge, as policy makers are still struggling to come to grips with the multiple effects of climate change, financial instability, new and resurgent infectious diseases, cyber-insecurity, international terrorism, and even war, to name but a few of the pending, unresolved global challenges. Yet further challenges loom on the political horizon, such as the specter of natural resource scarcities, rising resistance to antibiotics, and the implications that the future digital economy and society might have for production systems and labor markets worldwide.

The lengthening list of global challenges has not gone unnoticed. In fact, it has unleashed a flurry of palliative efforts aimed at risk prevention and management, including the strengthening of the resilience of individual households, local communities and countries, and firms and international supply chains. Moreover, innumerable corrective efforts have been initiated in most global challenge areas. However, they have not yet gone far enough and, evidently, they failed to allay people's fears that governments, individually and collectively, are no longer fully in control. Protest movements have arisen in the North, calling for greater concern about domestic human security (Stiglitz 2017; Reich 2016; Rodrik 2017). In other parts of the world, people are feeling compelled to flee due to war and lack of economic opportunities, leading to rising numbers of internally displaced persons, refugees, and illegal migrants (UNHCR 2017). And, as we also know, many people, especially women and children, are even too weak to escape from their misery and succumb to premature morbidity and mortality.

Certainly, conditions of human insecurity often stem from a variety of causes, whether personal and context-specific or local, regional, and national in origin. But global challenges also come into play; and many of these challenges affect us all, in rich and poor countries, powerful and fragile states alike. In fact, in many cases, problems of insecurity originate from an underprovision of global public goods (GPGs). For example, instead of climate stability we face global warming; instead of financial stability, excessive financial volatility; and instead of cyber-security, a rising number of cyber-attacks and blurring lines between "public" and "private."

Why do so many global challenges remain unresolved, despite the well-documented fact of their mounting costs?[1] Importantly, what could be done to tackle them more effectively and reduce the current threats to human security? And, could public policymakers not be confident that their constituencies would reward them for not only announcing but also fostering actual progress towards more sustainable and inclusive global growth and development?[2]

The present paper addresses these questions. The first section takes as its starting point the observation that many of today's global challenges originate from GPG underprovision and examines the key properties of these goods. Against this background, the second section identifies the current impediments to adequate GPG provision and possible corrective measures, including: fostering fairness of process and outcome in GPG-related international cooperation; building consensus on a notion of a mutually respectful exercise of national policymaking sovereignty; placing the GPGs themselves at the center of policymaking; and the creation of a Global Stewardship Council.

The main conclusion to draw from this analysis is that the root cause of the current manifold threats to human security lies in the ongoing, yet still unsettled transformation from the conventional Westphalian world order (WWO 1.0) to a more modern Westphalian world order (WWO 2.0) – that is, a system of global governance designed to allow policymakers better to combine global economic openness and interdependence among countries with national policymaking sovereignty. While this may sound like a daunting task, the four suggested reforms to advance this transformation involve relatively modest institutional adjustments. If the international community were to adopt reforms along these lines, human security worldwide might improve. The concluding section will address the question of why, although eminently doable, such corrective action is forthcoming only slowly.

Today's global challenges: Signals that global public goods are underprovided

Global challenges are called "global" because they often affect all people and all countries or, haphazardly, anyone anywhere on the globe. They thus bring to mind the economic concept of public goods. A closer look at these challenges reveals that, indeed, they are associated with public goods, resulting from an underprovision of GPGs. For example, instead of global financial stability we still experience – due, among other things, to inadequate financial-market regulation – the risk of excessive financial volatility. Likewise, instead of climate stability – due, among other things, to reluctant emissions reductions – we must contend with global warming, accompanied by more severe weather patterns like droughts, floods, and storms. For an adequate understanding of how to resolve such GPG-linked policy challenges and reduce the current threats to human security, it is thus useful to clarify what are the characteristics of public goods in general, and of GPGs in particular. (See Box 6.1, on the concept of human security.)

BOX 6.1 CONCEPTUALIZING HUMAN SECURITY

Human security is a condition that exists when people are free from perceived risks or threats that could endanger them, including threats that might prevent them from satisfying their most basic survival needs, leading a life of dignity, or enjoying relative stability in and control over their life chances and achievements.

Threats to human security – and thus experiences of insecurity – can arise at all levels of income and development. And, they can result from a wide range of global, regional, national, and local socio-cultural, economic, environmental, and political, as well as personal, factors or combinations of those.

Human security and insecurity can thus be viewed as potentially universal concerns. However, they may mean different things to different people and in different contexts, depending, among other things, on the source of the threats that are – or are feared as – posing risks to people's life and well-being.

Also, while some may view risks of change as potentially entailing danger, others may hope for change and prepare for the risks and opportunities that may come along with it.

Although the notion of human security has attracted growing attention during the past several decades, there exists as yet no consensus definition. The concept was introduced into the international debate by the *Human Development Report 1994* (UNDP 1994). For comprehensive overviews of the literature and the current state of the debate, see among others Acharya (in this volume), Fukuda-Parr & Messineo (2012); MacLean et al. ([2006] 2016); Tadjbakhsh & Chenoy (2007), and WBG (2013).

According to standard economic theory, public goods are non-rival and non-excludable in consumption. A good is non-rival if its consumption by one individual does not reduce its availability for other individuals; and it is non-excludable if it is technically unfeasible or economically undesirable to exclude additional individuals from its consumption. Goods that possess both these properties in full are said to be pure public goods, while goods that possess one or both properties only in part are called impure public goods. (The atmosphere is an example of an impure public good: rival in consumption, yet difficult to be made excludable.) Depending on the geographic reach of their effects, costs, and benefits, public goods are classified as local, national, regional, or global. Some public goods, including GPGs, may also be of inter-generational reach: that is, they generate long-lasting public effects, spanning several generations.

But it is especially important in the present context to clarify the meaning of "global." Some scholars employ the term "global" as a synonym for "international" or "transnational." However, as Kaul et al. (2016) argue, many GPGs are all of that and more. Because they have public effects of worldwide reach (as, for example, the atmosphere has), they not only extend across many or all countries and areas beyond national jurisdictions (such as the high seas), but they also penetrate into these countries and areas, whether invited in or not.

As stated in Box 6.2 and shown in Figure 6.1, many GPGs are not only global public in consumption but also global public in provision, meaning that in all or most countries, state and nonstate actors may have to contribute to their provision, even if only one or a few countries prefer to change the current design or level of a good's provision. GPGs of this type thus entail policy interdependence among countries. Or, put differently, they compel countries to seek the political support and cooperation of others, if they wish either to modify the provision level or form of an existing GPG or to create a new one, such as the R2P principle.[3]

BOX 6.2 DEFINING GLOBAL PUBLIC GOODS

Global public goods (GPGs) share with other public goods (PGs) the key property of publicness in consumption: being fully or partially non-rival and non-excludable. What distinguishes them from other PGs is the reach of their publicness in consumption, which (1) spans several geographic regions or even the globe as a whole; (2) may penetrate into countries, areas beyond national jurisdictions, or both, with variable levels of impact; and (3) may be of long-term duration, affecting, for better or worse, several generations. Thus, while criterion (1) is the prerequisite for a good to be defined as a GPG, the publicness in consumption of GPGs could potentially comprise three dimensions:

- A *spatial* dimension: being of worldwide scope;
- An *impact* dimension: affecting countries and areas beyond national jurisdiction;
- A *temporal* dimension: having long-term effects.

In most cases, global publicness in consumption along any of these three dimensions will not be an innate property of the good but rather will reflect a policy choice or the lack thereof.

In addition to being public in consumption, many GPGs, like other PGs, are also public in provision: Their provision involves a large number of actors and compels countries to seek the cooperation of others.

Source: *Kaul et al. (2016)*

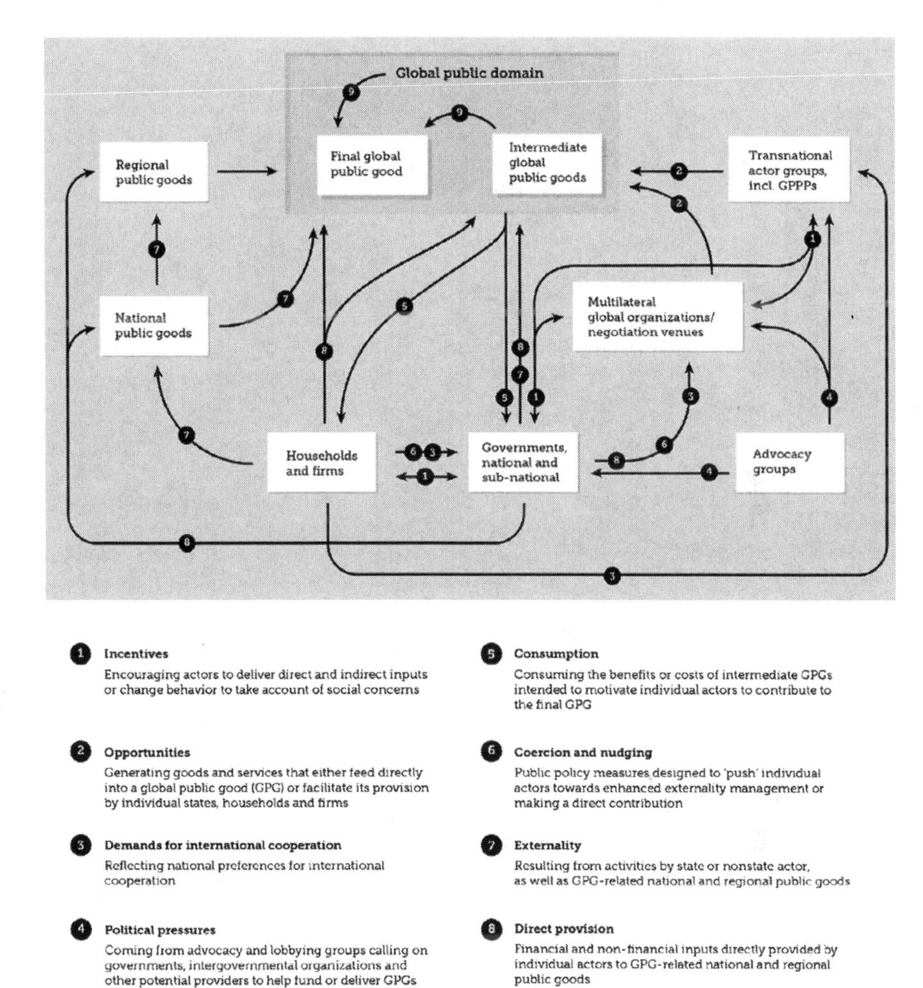

FIGURE 6.1 Provision of a global public good
Source: Kaul et al. (2016), based on Kaul & Conceição (2006: 14)

Moreover, the publicness (or privateness) of a good is often not an innate property of it but a social construct. The same holds for its geographic reach, including the global publicness in consumption of goods. Some GPGs, notably the global natural commons (such as the moonlight and the ocean), have always existed; they are natural GPGs. Over time, however, and particularly over the past half-century, an increasing number of human-made GPGs have been generated, partly to facilitate globalization and partly as a result of that phenomenon. For example, norms pertaining to intellectual property rights or basic human rights were intended to become global public goods in consumption accepted by all, and therefore have been actively promoted. Yet issues like communicable diseases unintentionally have assumed stronger GPG relevance in the wake of globalization, as the greater openness of national borders and a rising volume of cross-border economic activity have allowed them to move with ever greater ease and speed across the world.

Clearly, GPGs are complex issues. However, in most instances, their scientific and technical aspects are well understood by now as are many of the policy approaches and instruments that could be employed to enhance their provision (see Kaul, Blondin, & Nahtigal 2016). So, what are the factors that currently constrain policymaking and hold back a more effective resolution of global challenges? And what possible reforms might be adopted to overcome the present impediments?

Fostering adequate GPG provision: Combining effective international cooperation *and* national policymaking sovereignty

The current underprovision of GPGs is likely caused by a large number of diverse factors. In addition, the exact combination of impediments is likely to vary from one type of good to another. But, as the purpose of this paper is to examine why, in today's world, GPG-type challenges across-the-board appear to be difficult-to-govern issues, the following discussion focuses on an aspect that most GPGs share, viz. the structure and functioning of the present governance systems at national and international levels. Of special interest is to examine whether these systems need to be modified and, if so, what could be done to equip them better for addressing GPG-related challenges under the current policymaking realities.

Assuming that the goal is to promote more sustainable and inclusive global growth and development and, with it, enhanced human security, the following four reforms appear to warrant attention: (1) fostering fairness of process and outcome in GPG-related international cooperation; (2) building consensus on a notion of a mutually respectful exercise of national policymaking sovereignty; (3) placing the GPGs themselves at the center of policymaking; and (4) the creation of a global stewardship council.

Reform 1: Fostering fairness in international cooperation

We are living today in an increasingly multi-polar world or, as Acharya says, in a "multiplex world" characterized by five main features: absence of global hegemony; proliferation of major actors; complex global and regional interdependence; multi-level governance; and multiple modernities, that is, a "world of cultural, ideological, and political diversity, including alternative ideas [about] pathways to stability, peace and prosperity" (Acharya 2017d: 11). This multiplex nature of today's world is no surprise, considering the vast global differences and disparities that still exist, despite all the convergence that has happened. And, therefore, it is also not surprising that, in many cases, people's and countries' preferences for certain GPGs vary, including on how to provide and shape them and how to distribute related costs and benefits. With increasing multi polarity and the proliferation of state and non-state actors involved in international cooperation, calls for "voice reform" – for affording all the concerned parties an effective say in matters that concern them – have also multiplied. But so far, they have been met only to a limited extent (cf. Wade 2014, Weisbrot & Johnston 2016, Reisen & Zattler 2016). In part, the response has even been a retreat, notably on the part of the more powerful states, from the universal international forums into more mini-lateral or "club" arrangements, at both the policymaking and the operational or implementation level. Thus, at a time when it is important for "all hands to be on deck," international cooperation has been increasingly fractured and disjointed.

A possible corrective step could be the one depicted in Figure 6.2: to align the publicness in decision-making on a particular GPG to its publicness in consumption and provision so that all the concerned parties are being consulted and can have an effective say, nationally and internationally. Doing so would increase the likelihood that the provision process will lead to a mutually beneficial outcome acceptable to

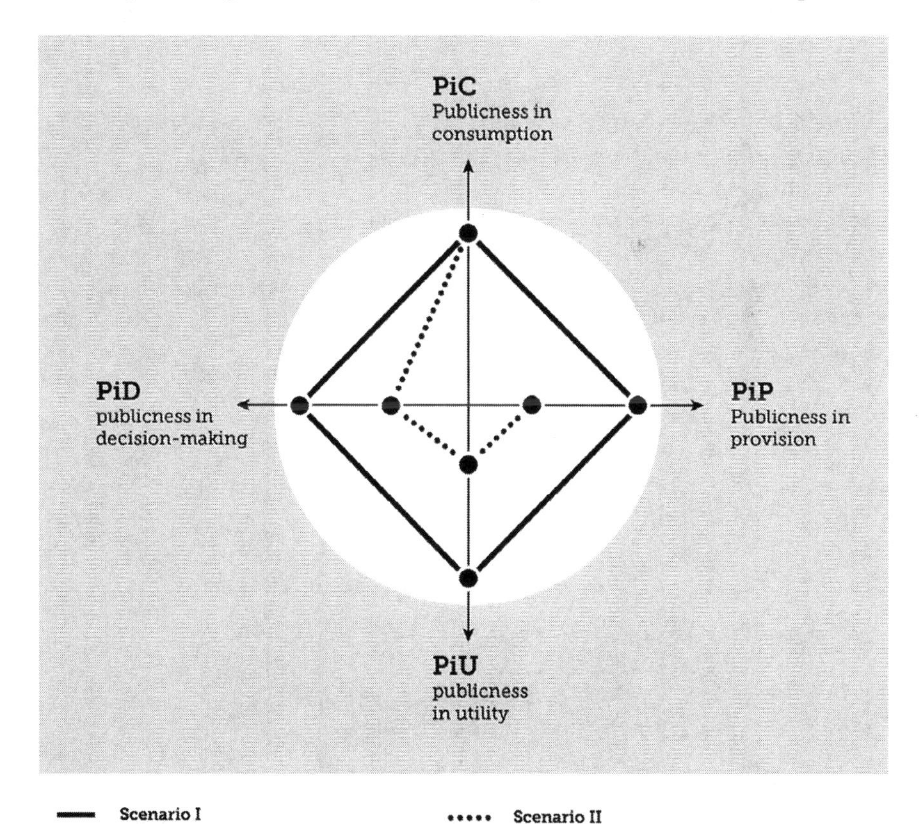

FIGURE 6.2 The four dimensions of publicness
Source: Kaul et al. (2016)

all parties, issuing in what Figure 6.2 calls "publicness in utility." Such fairness of process (decision-making) and outcome (a fair distribution of costs and benefits) might help strengthen actors' willingness to cooperate, so that the good's requirement of global publicness in provision would be fully met.

Reform 2: Building consensus on a mutually respectful exercise of national policymaking sovereignty

Inviolability of national borders and non-interference by outside forces in the domestic affairs of states are key principles underlying the present world order. Yet, as discussed above, GPGs often enter into national policy domains whether they are welcome or not. They run counter to the fundamental principle of the present world order; thus, they are likely to be contested issues when states perceive them as adversely affecting national welfare and well-being. Controversy may erupt especially when preferences for GPGs vary among countries and population groups that exercise different degrees of political influence. As noted previously, this is often the case.

Therefore, without voice reform and credible arrangements for more fairness in international cooperation, many countries may (often for good reasons) view GPGs and related international cooperation initiatives as undermining their national policymaking sovereignty, especially when driven by top-down power politics. Yet, non-cooperation leaves GPGs underprovided, ultimately making all countries worse off and leading to a condition that Kaul (2013: 33–34) has termed the "sovereignty paradox." While shying away from international cooperation to safe-guard their policymaking sovereignty, states achieve the opposite of what they intended: more and more crises, which cumulatively undermine their policymaking sovereignty instead of strengthening it.

A way out of this policy trap would be the reform step outlined in Box 6.3, viz. for the United Nations to launch an initiative to encourage states to agree to a reconceptualization of national policymaking sovereignty that would allow them to better combine openness and connectivity with the pursuit of national interests.

BOX 6.3 A RESPONSIBLE EXERCISE OF NATIONAL POLICYMAKING SOVEREIGNTY: A COLLECTIVE WAY OF SAFEGUARDING NATIONAL POLICYMAKING SOVEREIGNTY?

International cooperation is often seen as undermining states' policymaking sovereignty. No doubt, it often does; and therefore, governments frequently shy away from a global, concerted policy response, even in issue areas that involve transnational challenges which no single nation can effectively and efficiently address alone. In the absence of a cooperative approach, global challenges will linger unresolved, potentially making all parties worse off.

Thus, when confronting challenges that entail policy interdependence, it is often in the enlightened self-interest of all concerned states to offer fair and mutually beneficial cooperation. This requires mutual confidence and trust. Accordingly, there must be a shared commitment among states to act responsibly, both toward their own territories and constituencies – protecting against negative spill-ins from abroad – and toward other states, because non-cooperation could undermine welfare and well-being for all.

In other words, exercising responsible sovereignty means pursuing national interests in a way that is fully respectful of both the sovereignty of other nations and the systemic integrity requirements of GPGs and, to that end, oriented toward the maintenance of global balances and planetary environmental boundaries.

Just as states' commitment to the norm of collective security strengthens the inviolability of national territorial borders, a commitment to exercising their policymaking sovereignty in a mutually respectful and responsible manner could be, in areas of policy interdependence, the best way to secure their national policymaking capacity – provided that international-level decision-making on global challenges is marked by process and outcome fairness.

Sources: *Based on Kaul (2013) and Kaul & Blondin (2016)*

The suggested notion of a mutually respectful exercise of national policymaking sovereignty could imply that, in GPG-related policy fields involving interdependence and cooperation among countries, states agree not to resist cooperating with others, provided that the agreed-upon arrangements are, on the whole, mutually beneficial. Such willingness to cooperate would also entail another commitment: wherever possible – and in line with such established global principles as that of common but differentiated responsibilities and capacities[4] – states should aim to achieve enhanced management of cross-border spillovers, especially spillovers that might adversely affect other countries directly or indirectly, whether by diminishing GPGs or harming areas beyond national jurisdictions, such as the high seas. Some examples include the emission of greenhouse gases; non-reporting of the outbreak of a communicable disease; and the circulation of so-called toxic financial products.[5]

Several spillovers that are known to entail global social costs (that is, to harm other countries and the natural environment) now plague the global public domain, while others that could potentially accelerate progress towards agreed-upon goals like poverty reduction or climate change mitigation often are held back. For example, knowledge is a non-rival good. Yet, much of it is protected by patents and often is unaffordable to large segments of the world's population (cf. Stiglitz 2014).

Clearly, the management of externalities today is not working well. The reason often is that states, or more precisely their governments, maintain a conventionally strict notion of policymaking sovereignty. Forging global consensus based on the principle of a respectful and mutually beneficial exercise of national policymaking sovereignty thus could be a major advance towards reducing the world's current susceptibility to crises.

Reform 3: Placing the GPGs themselves at the center of policymaking

Judging from the scientific and technical literatures on various GPGs, it appears that many of these goods have their own systemic integrity requirements. As long as those are not maintained or achieved, they may fail to function in an intact and undiminished way. This holds for climate change mitigation as well as for financial stability or the eradication of polio.[6] But past experience has shown that these systemic integrity requirements of GPGs often exceed what state and non-state actors, individually and collectively, are willing to do to meet them. In many cases, countries contribute to GPGs to the extent that national and global interests overlap. The problem is that they often only partially overlap, so that the good in question continues to linger in a state of underprovision.

A possible remedy could be to introduce "global issue facilitation" as a new function into governance systems, nationally and internationally. In addition, one could prepare "provision path sketches" for each GPG in question (along the lines shown in Figure 6.1, but in greater detail). Such sketches would allow policymakers more effectively to monitor and assess the progress being realized towards the goal of adequate provision and, where shortfalls or delays occur, to identify and establish requisite incentive measures to strengthen actors' willingness to contribute.

If well-founded, such provision path sketches could also demonstrate that international cooperation in support of the GPG being addressed promises to be a relatively good investment and truly "pays." Nevertheless, it might be useful also to consider the following step, as the role of states vis-à-vis GPGs differs starkly from their role toward national public goods.

Reform 4: Creating a Global Stewardship Council

Within the domestic context, states often play an active role in public-good provision. They are expected to intervene where markets fail adequately to provide public goods or internalize externalities. Hence, they are usually also endowed with special coercive powers, including the power to tax. They can potentially prevent national public goods from becoming trapped in unresolvable collective-action problems. With GPGs, in contrast, states – for various reasons, including resource and capacity constraints or varying preferences and lack of political will – may be non-cooperators themselves, with no higher-level international authority (besides the threats originating from GPG underprovision) in place to compel or nudge them into concerted action. This is the problem to which Nordhaus's (2015) "Westphalian dilemma" refers. The policy concern then becomes how to escape from this dilemma.

Certainly, the availability of more well-founded evidence that, in GPG policy spaces (in which conditions of policy interdependence exist), it often pays to cooperate could perhaps help to persuade states to exercise their sovereignty in a considerate, respectful, and responsible manner. Such evidence-based cooperation thus would be one way of addressing the Westphalian dilemma. Another way – perhaps an important complement of the former – would be the creation of a Global Stewardship Council, possibly under the umbrella of the United Nations.

Kaul and Blondin (2016: 13–14) envision this council as "a standing body of eminent personalities, wise men and women, appointed in their individual capacities… not guided by particular interests, but by what is good for the global public." Each of its members would represent a particular GPG, including, for example, the atmosphere, the ocean, the global knowledge stock, financial stability, health, equity, peace and security, or any other GPG that, at a given point in time, might be viewed as warranting special attention. As the present world order also possesses the properties of a GPG,[7] it too could have representatives on the council, perhaps one each for the low-, middle-, and high-income countries. And the often-voiceless future generations should, no doubt, also be represented, as shown in Figure 6.3.

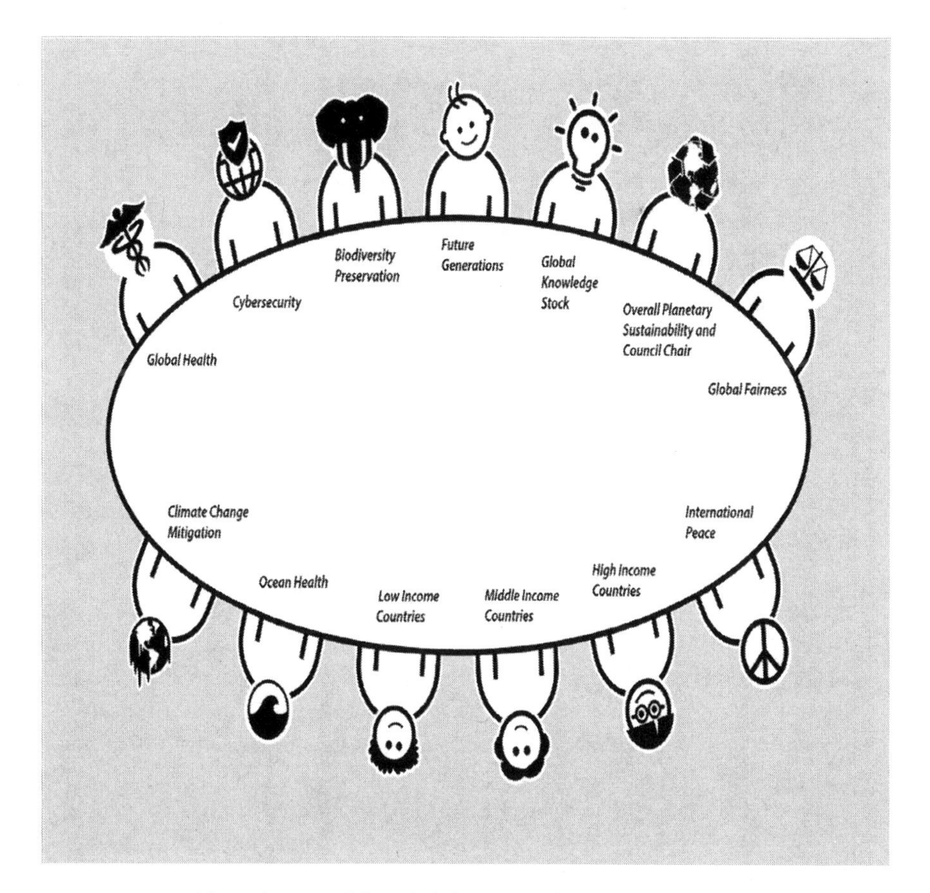

FIGURE 6.3 Possible make-up of the Global Stewardship Council

The council would have an advisory role but no legislative powers. It could point states towards policy paths and strategies that could make all better off, combining efficient, fair, and effective international cooperation for longer-term sustainability with respect for state sovereignty. As also noted in Box 6.4, a body of this type thus would not just be another forum in which states negotiate agreements based primarily on particular national (and human-centered) concerns, but one that reminds us that achieving global sustainable growth and development calls for accommodating both imperatives: the satisfaction of particular human and national interests and the requirements of adequate GPG provision, notably those of the global natural commons, where overutilization and underprovision could lead to irreversible harm.

BOX 6.4 THE ROLE OF A GLOBAL STEWARDSHIP COUNCIL

Past experience has shown that, when participating in international negotiations, states are often primarily concerned about inserting their respective national interests into the debate rather than about reaching consensus on how to meet the systemic integrity requirements of the global public good (GPG) under consideration and ensure its adequate provision. In many instances, including in the climate field (where critical thresholds of potentially irreversible harm are fast approaching), states' willingness to take corrective action goes only as far as national interests overlap with global interests. But this is often not the case; and therefore many GPGs remain underprovided and global challenges unresolved.

Thus, it might be worthwhile to consider the creation of an independent Global Stewardship Council. The members of such a council could be eminent personalities, selected through a worldwide, open, and participatory process, and nominated in their personal capacity as representatives of (1) certain GPGs, including the atmosphere, the ocean, financial stability, global equity, the global knowledge stock, and global agreed-upon norms such as the basic human rights; and (2) certain groups of states and the future generations.

Individual council members would act as advocate of the issue or group they represent, while taking global interests and exigencies into account; and when discussing global policy they would remind fellow members of the special needs that their issue or group might have. Their main goal would be to identify, based on agreed-upon principles of fairness, possible ways to promote global sustainable and inclusive global growth and development. Wherever possible, they would aim at suggesting mutually beneficial bargains that might persuade state and nonstate actors to take faster and more decisive action on unresolved challenges.

The Global Stewardship Council best could be located within the United Nations. But, during an initial pilot phase, it could also be hosted by an entity such as the World Economic Forum.

The creation of such a Global Stewardship Council would not be a totally new invention, because similar types of bodies exist already. For example, we might mention the Global Futures Councils established by the World Economic Forum,[8] the World Future Council,[9] and the national councils for sustainable development (NCSDs). Their function, too, is to think about the longer term and about policy measures that the respective states could promote domestically and abroad to foster longer-term sustainability. NCSDs – often multi-stakeholder bodies that review, and advise governments on, such matters – already exist in about forty countries (Osborn et al. 2014). Evidently, the need for institutional innovation to support sustainable and inclusive global growth and development is being felt. The Global Stewardship Council could be an important advance in creating a global (multi-level, multi-actor) governance architecture suited to this purpose.

If implemented together, the aforementioned four reform options might make it possible for openness and policy interdependence to be combined with national policymaking sovereignty, thus helping to usher the world into a new era: a Westphalian order 2.0. It would still be a based on sovereign nation-states, but with the addition of a new multilateralism that recognizes multipolarity and acknowledges that the best way to meet national interests often is for states to engage in effective and fair international cooperation, at least in GPG-related policy fields.

Conclusion: If it's doable, why ain't it been done yet?

The present article has explored the reasons why so many GPGs today are under-provided, allowing excessive volatility, disaster, deprivation, and violence to threaten human security worldwide. An important reason could be that corrective measures have been slow in coming, and that, instead, the policy responses to date mainly have emphasized the strengthening of communities' resilience by helping them to absorb external shocks and improve their risk management (cf. Hartwig & Wilkinson 2016, OECD 2014, WBG 2016). New insurance products have been developed, including (for example) insurance against cyber-attacks, terrorism, and other large-scale catastrophes; meanwhile, similar products, such as insurance against ill-health or natural disasters, have been redesigned so they will become more affordable (Shiller 2013). Protection against threats to human security has turned into a flourishing business.

Clearly, the more affluent population groups can afford to buy private protection. They can live in gated communities, purchase comprehensive insurance, or fly to work by helicopter. Then too, the richer countries can make their own national arrangements. Yes, in many instances human insecurity is indeed global-public in its effects, but because some people more easily can afford protection against it than others, political support for getting to the root of the problem may not yet be sufficiently strong.

But protection against the threats of human insecurity will only go so far. As many of the causes are global-public, so must be the solutions. Therefore, the question that must be asked is: how much worse must things get before we, the global public, and our policymakers will be prepared to take the step forward into the WWO 2.0, creating a new modernity of "enhanced human security for all"?

As the present discussion has shown, the new multilateralism underpinning WWO 2.0 not only would reduce the current "crisis is the new normal" condition of the world but also (re-)strengthen the policymaking sovereignty of states, citizens, and governments. Then one might find that laying the foundation for the shared future of enhanced human security opens up important new avenues for a world of multiple modernities: people having more opportunities and greater capabilities to live lives of their choosing.

Notes

1 See, for example, on the threats to human security from climate change, IPCC (2014a) and (2014b: 755–91); and on the economic costs of climate change, IPCC (2014c: 15–17). On global public health threats in the 21st century, see WHO (2007), and on the human and economic costs of one of the challenges likely to confront us, viz. antimicrobial resistance, see Leatherby (2017). Estimates of the mounting global cost of cybercrime are, among others, presented in CSIS (2014). See Agrafiotis et al. (2016) on concepts, taxonomy, and measurement of cyber harm, including harm to a variety of stakeholders. On the impact of banking crises on people's lives, see WBG (2013, notably chapter 6). And, on the costs to society of the barriers still constraining women's opportunities (and, thus, signaling under provision – incomplete universalization – of the norm of gender equity), see, for example, Ferrant & Kolev (2016), ILO (2017) and Klugman et al. (2014).

2 For examples of international agreements proclaiming such goals, see the two landmark agreements of 2015: *Agenda 2030* and the *Paris Agreement. Agenda 2030* was adopted by world leaders at a special summit meeting of the United Nations. It sets forth 17 main goals and 169 sub-goals that the international community aims at achieving by 2030. The goals apply to both developed and developing countries, with the overarching objective being to foster inclusive and sustainable growth and development worldwide. For details, see www.un.org/sustainabledevelopment/development-agenda/. The *Paris Agreement* is the outcome of the 21st session of the Conference of the Parties of the United Nations Framework Convention on Climate Change (UNFCCC). Its central aim is to strengthen the global response to the threat of climate change. For details, see http://unfccc.int/paris_agreement/items/9485.php/.

3 The principle of responsibility to protect – also known as R2P – refers to the obligation of every state to protect its populations against mass atrocities and to the responsibility of the international community to assist individual states in meeting that obligation. However, if a state manifestly fails to protect its population, the international community must be prepared to take appropriate collective action. In its resolution A/RES/63/308, adopted in 2005, the United Nations General Assembly endorsed the R2P by consensus. See www.un.org/en/genocideprevention/about-responsibility-to-protect.html/. For the debates preceding and following this resolution, see Acharya (2013a) and Thakur (2017), among others.

4 As stated in the 1992 United Nations Framework Convention on Climate Change, the principle of common but differentiated responsibility means that "[t]he Parties [to the Convention] should protect the climate system for the benefit of present and future generations of humankind, on the basis of equity and in accordance with their common but differentiated responsibilities and respective capabilities. Accordingly, the developed country Parties should take the lead in combating climate change and the adverse effects

thereof." In addition, the Parties should take account of "their specific national and regional development priorities, objectives and circumstances." The Convention is available at https://unfccc.int/resource/docs/convkp/conveng.pdf/.

5 The Financial Times Lexicon defines toxic assets as assets "such as the securitization of subprime mortgages, where the original creators of the securities failed to take into account the real rate of mortgage default and the extent to which it would be contagious across securities." As a result, AAA-rated assets could suddenly look like junk bonds that no one wants to buy. Available at http://lexicon.ft.com/Term?term=toxic-assets&mhq5j=e1.

6 But there are also GPGs that can be incrementally provided. For example, global norms such as the basic human rights have been – and continue to be – globalized in a gradual and, not uncommonly, recurrent manner.

7 In order for a world order to be viewed as legitimate and worth maintaining, its basic underpinning principles must be global public in consumption and, preferably, must be respected voluntarily.

8 The mandate of the Forum's Global Futures Councils is to assess "the future of the most important systems that shape global transformations, with clear recommendations for our actions today" and to provide "analysis on the future impact of the Fourth Industrial Revolution on different systems [and] innovative ideas, frameworks, processes and recommendations for the effective, value-based governance of those technologies." For details, see www.weforum.org/communities/global-future-councils/.

9 The World Future Council is an independent, non-profit organization headquartered in Hamburg. Its purpose is to undertake research on future-just legislation and to advise and support decision-makers in the concrete implementation. See www.worldfuturecouncil.org/about/.

PART IV

Dialogue

7

THE PAST AND PRESENT OF EUROPE'S INTERCULTURAL DIALOGUE

Beyond a "normative power" approach to two-way cooperation*

Mario Telò

Over the course of its long history, the European theory and practice of intercultural dialogue with other civilizations has developed along many paths, and displayed numerous forms and degrees of effectiveness. However, it has been less easy for Europeans to recognize the existence of multiple modernities, distinguish between modernization and Westernization, and establish an open and equal dialogue with the "others." For many historical and geographical reasons, Europe has been interacting with the "others" since its first attempts at unification in ancient Roman times. In the process, it has tried out various forms of communication ranging from inclusiveness and a genuine interest in understanding the other to Eurocentrism and the will to conquer. At times, it has managed to carry on communication as a two-way street.

The following essay will address this essential topic by taking the reader on a journey through several historical cases, showing both the promise and the limits of the Europeans' successive interactions and dialogues with other civilizations. The journey will pass from ancient Rome to the first wave of globalization in the 16th and 17th centuries, culminating in the position of the EU today as an unprecedented kind of global power within a globalized and multipolar world. We will keep in mind that "becoming conscious of the relativity and arbitrariness of the segments of our history already means changing it to some extent" (Todorov 1992).

The first paradoxical example: Ancient Roman citizenship

Although the Greeks – in the person of Socrates – invented the concept of the "universal" in the 5th century BCE, they also branded as "barbarians" all those peoples who were not fluent in speaking the Greek language. By contrast, the Romans dealt with "others" in a twofold manner: They distinguished the members of the Roman Republic (and later Empire) in its broadest sense from unassimilated outsiders beyond Rome's borders. The internal challenge to Roman thinking and institutions should not be underestimated; it was daunting in scope and intercultural in nature. In fact, Rome's citizenship policies are the first example of regulated globalization in European history. Those policies evolved as the polity grew from a relatively small city-state to a power that dominated the Italian peninsula, until it matured into a multi-continental Empire. During the five centuries of the Republic, Rome assimilated the Latin inhabitants of the Italian peninsula. During the next five, when it had become an Empire after Julius and Augustus Caesar, Rome incorporated all of the peoples living within the far borders (*Limes*) of the Empire. How did the Romans integrate the others? According to international experts (e.g., Rémy Brague 1992), the best example of the Roman development of the Greek cultural legacy was the invention of the legal notion of citizenship as a status that should not be limited exclusively to the residents of the city of Rome. Eventually, the status of "*Civis Romanus*" could be conferred upon all free inhabitants of the Empire without discrimination according to religion, ethnicity, or language.

Building on this open citizenship policy, which originated in 46 BCE (when citizenship was expanded from the city of Rome to the inhabitants of today's northern Italy), Caracalla's Emperor Edict (212 CE) conceded Roman citizenship to all inhabitants of the Empire, from Hadrian's Wall (the border of Scotland) to the borders of the Persian Empire, from Lusitania to Egypt, from Dacia to the current Germany. All distinctions between conquerors and defeated populations or between people living in the cities and in the countryside were legally abolished. Newly created Roman citizens could now benefit from the Roman written private civil and criminal law, as well as enjoying civil and political rights, whether passive or active. As a consequence, during the imperial era, high-ranking civil servants, consuls, and even emperors might be born in provinces outside of Italy (Trajan and Hadrian, for example). The word "racism" does not exist in Latin. The conferral of citizenship was a very efficient way to attain two objectives: consolidation of stability within the borders and active integration of a multicultural political entity, making uprisings, riots, or internal revolts against the central government quite rare for more than two centuries. Within the bounds of its *Limes*, Rome provides the first example of legally framed globalization.

However, a time of troubles began after 378–380 CE, when large-scale immigrations/invasions from the East began to overwhelm the Roman polity (Barbero 2007). By tradition, Rome ruled the populations of its multi-ethnic Empire by pluralist policies and by tolerating diverse religions, a principle that was confirmed by the Edict of Milan of 313. But, it also employed two other tools of integration: Citizens from non-Roman ethnicities could either be drafted as soldiers into the army, or, if they were peasants, they could be integrated by distributing agricultural land to them. All of these policies began to show their limitations, and the Western imperial government became less and less able to govern a society that had become increasingly complex demographically. On the one hand, the Emperor Theodosius declared Christianity to be the sole religion of the Empire (Edict of Thessalonica, 380 CE) merging for the first time citizenship and religion; on the other hand, the inclusion of the Goths within the army looks like a colossal blunder after a series of tragic incidents between the army and resident populations in the Empire. These events marked the beginning of the internal disintegration of the Empire, at least of the Western side, because the Eastern Empire lived on as a relatively inclusive, multicultural polity until 1453.

The second complementary feature of the Roman Empire was the absence or perfunctoriness of dialogue with outsiders – those who lived beyond the *Limes* – who did not even enjoy trade relations with Rome. The Empire was never conceived as universal in its aspirations, in contrast to the "Catholic" (universal) religion that eventually became its official creed. It always featured geographical boundaries and an ambiguous relationship with the outsider-others. The economic objective of the Empire was autarky, a far cry from (say) the trade policies of the Dutch, British, and even American global empires. The *Limes* frequently underwent changes in its length, extent, and distance from Rome, as well as in its function, whether offensive or defensive; however, it demarcated the insiders (whatever their ethnic and racial characters), as potential citizens, from the outsiders, who had not (yet) joined the Roman Empire. Throughout successive works of construction and consolidation, the *Limes* became a security zone surrounding the Empire, from the Middle East to the Danube, to Germany and, on the South, throughout the northern coast of Africa. Only the Great Wall of China, constructed as a consequence of the parallel process of unification of China under the first Emperor Qin Shi Huangdi (Qin dynasty, 221–206 BCE) and later the Han dynasty was comparable in extent and intent, but even it was less developed in geographical terms.

The contradiction between internal integration within the Empire and exclusion of the outsiders (those living beyond the *Limes*) has been addressed by recent research (Horchani & Zolo 2005), which has focused on one of the most challenging borders of the Roman Empire, namely the Mediterranean, and contrasts the concept of *Limes* (exclusive) to the concept of *Limen* (inclusive). Following Braudel's pioneering work on the common Mediterranean civilization (1949) and the studies of Beck, Bauman, and others on globalization, the authors suggest that Roman culture itself offers the conceptual tools for distinguishing between *Limes* and *Limen*, where *Limen* means threshold, beginning, passage, and – as a consequence – openness, pluriversality, tolerant communication, encounters with outsiders. This Latin concept provides the background of the search for commonalities, of *open texture*: "*Distinguer pour unir*" means looking for a relationship precisely by taking distinctions into account.

European relations with the Other during the first wave of globalization: Alternative perceptions and policies

Contrary to the universalist commitment announced by its self-definition as "Catholic," European Christian identity during the *Respublica Christiana* of the Middle Ages was *de facto* inward-looking, exclusive, and intolerant in respect to its relationships with the two "others" it knew best: the Jews (victims of persecutions and expulsions) and the Muslim Arabs (enemies of the Crusades), and that in spite of the mutual intensive cultural interaction. The European identity of those times was rather a "wall identity" (Cerutti et al. 2010), a construction against the others, at least until historical change occurred with the birth of European modernity. The historical context changed dramatically with the scientific revolution, the development of the internal market, the consolidation of the first modern states and – last but not least – with the first wave of globalization, starting at the end of the 15th century.

While losing its internal religious unity after the Protestant Reformation and subsequent Wars of Religion, Europe in the era of Habsburg Emperors Charles V and Philip II became a global actor. It was economically and technologically on a par with several other world power centers (Kennedy 1985): the Ming dynasty in China, the Moghul Empire of al-Akbar in India, the State of the Grand Prince of Moscow (later on known as Tsarist Russia), and the Ottoman Empire. However, beginning in the 15th and 16th centuries, Europe proved able to exert varying forms of influence on the other continents more than any other civilization had been able to do previously. In this respect, the end of the 15th century, with the discovery of the Americas and the intensification of contacts with Asia and Africa, does represent a critical juncture, a true historical turning point.

European civilization, including the process of modernization and industrialization, was exported overseas by many means, sometimes relying on soft power but also employing imperial domination and violence. That civilization eventually was rejected or emulated by many "others." We will focus on two case studies, those of the Americas and China, keeping in mind the threefold distinction proposed by Todorov as possible types of relationship to the other: values appreciation, kind of relationship, and knowledge.

Understanding the Other in the case of the Americas: Eurocentrism, colonialism, philanthropy, and relativism

The first wave of globalization after the Western voyages of discovery made by Columbus, Vespucci, Verrazzano, and others brought to Spanish America the conquistadores, the Dominicans, and the Franciscans, while in the Portuguese sphere of influence, Vasco da Gama and the admirals and Jesuits who succeeded him established relations with both Native Americans and Asians. These voyages created an extraordinary context for a new debate about the relationship of Europe with the other.

The discovery of the American continent implies two new challenges for the Europeans: first, understanding and interacting with the Indians; second, reconstructing a new global synthetic theory of the world and the place of Europe within it. According to the French theorist of the modern state, Jean Bodin, "*[T]ous les hommes sont reliés entre eux et participent merveilleusement à la République universelle comme s'ils ne formaient qu'une seule et même cité*" (Bodin 1566: 298). What new synthesis between the European aim of universality and the evidence of increasing diversity would emerge? What common core of all the different types of civilizations could be identified?

The problem of understanding the Native Americans and the New World inaugurated a major and multi-faceted intellectual controversy about otherness. In contrast to the approach of Rousseau, for whom "*Les sauvages sont les plus parfaits représentants du genre humanaine,*" G. L. Buffon (1761) and Cornelius de Pauw (1768) initiated a debate on the intellectual and moral status of native peoples. They considered various arguments supporting the thesis that Indians were naturally inferior, ranging from out-and-out racism to alleged zoological laws, natural history, and climate theory. Most of these pictured the Indians as similar to animals, but also as degenerate, corrupt, and weak. Accordingly, these scholars added policy recommendations condemning as useless most pleas in favor of protecting and helping Native Americans, while justifying the exploitation of peoples defined as "slaves by nature."

These "theories," justifying despicable practices widespread in the history of Western European imperial policies such as massacres of innocent people, reveal the collective responsibility not only of Spanish conquistadores like Cortés (as Las Casas asserted in his last books), but also of Portuguese, Dutch, English, French, and later North American colonizers. Their actions, and the theories that supported them, have provoked an intellectual controversy that has focused first and foremost on values (Gerbi 2000).

Obviously, Europeans took for granted the superiority of their own values. That assessment carried over into other fields as well, including what passed for knowledge and the influence of that knowledge on practical relationships, often instrumental to the harshest forms of political and economic domination. In G. W. F. Hegel's dialectical *History of Philosophy* one finds, for example, a humiliating and cruel role assigned to the "defeated peoples" of the Americas. Yet one also notes the potential for a "future American rationality beyond Europe."

Among the relatively wide range of relevant opponents of this vision and practices, from Rousseau to Humboldt, a special place has to be given to Bartolomé de Las Casas (1992). The evolution of the Las Casas approach in favor of protecting the Indians still looks particularly interesting. Writing much earlier than they, Las Casas challenged, by anticipation, many of the key arguments of Buffon, de Pauw, Sepulveda, and others. Their thesis, we recall, had been that the Indians were inferior, especially as evidenced by the barbarity of practices such as human sacrifice. Las Casas argues that every civilization has allowed human sacrifice (the Greeks with Agamemnon and Iphigenia, the Old Testament Jews with Abraham and Isaac, the New Testament Christians with God the Father and Jesus Christ). Later, in his *Apologética Historia,* he moves from the thesis of evolutionary development and similarity (which could function as background assumptions for assimilation policies) to the argument in favor of "relativism" (defined in this context as similar to "perspectivism") as far as the ways to God are concerned: "each one is barbarian to the other." However, in his mind, understanding the others and adapting to their ways should not be merely a prelude to dominating them, as it had been in the

Cortés approach, but the opposite. The political implication of Las Casas' viewpoint in the *Apologética* is that the conquistadores should respect the autonomy of the Indians. His emphasis on relativism is the Christian, tolerant, and pluralistic answer to the question raised by Jean Bodin about the future of the "universal republic."

However, the majority of contemporary European observers considered the Americas either as a kind of immature object of history or as a continuation ("Annexum" by Hegel) or expansion of European civilization. By contrast, ever since ancient times Asian civilizations (China, Persia, and India) have been considered to be a more challenging other, truly representing alternative worlds.

China as the symbol of the Other ancient civilization: The European debate of the 18th century

It is impressive that a century later, during the *Siècle des Lumières*, understanding China was destined to become one of the most relevant topics of the European philosophical discussion about multiple modernities. That debate engaged the brightest minds of the day: Helvétius, d'Holbach, Voltaire, Diderot, Rousseau, and Montesquieu, among others (Van Staen 2016). The very broad spectrum of viewpoints ranges from an idealized picture of China as the most ancient and sophisticated civilization on earth, to contempt for a people that was conquered without much difficulties by the Manchus (then confused with another neighboring people, the Turkic Tatars or Tartars) in 1644.

The most balanced and objective approach was taken by Montesquieu. In terms of values, the dismissive approach to China is strongly argued, even if in quite different ways, by d'Holbach and Rousseau. D'Holbach opposes civilized and rational Europe to China, a country supposedly characterized by superstitions, traditionalism, Emperor-idolatry, and (consequently) despotism. In short, for those philosophers China provides an example of anti-modernity. Rousseau's critique of China is based on the opposite interpretation of events. The successful 1644 invasion of China by the less civilized Manchus offers evidence, according to the "citizen of Geneva," of the inevitable decline of all increasingly sophisticated, educated, and advanced civilizations. According to Rousseau, China provides the example of the weaknesses and corruption endemic to the model of civilization so much emphasized by Voltaire and the Paris Encyclopedia milieu. "I've seen the most illustrious and numerous nation of the world dominated by a few gangsters. I've seen this famous people, and was not surprised to find it a slave, conquered every time it was attacked, victim of every conqueror, and forever" (cited in Van Staen 2016: 105). The idea that China lost wars against conquering neighbors because of its excessive civilization was totally unacceptable to Voltaire, especially after the break between the two intellectuals became open in 1755.

In his "*L'Orphelin de la Chine*" Voltaire offers the opposite argument, using the interpretation of China's weakness as a weapon in his theoretical quarrel with Rousseau. He admires China more than Europe as an example of civilization defeating barbarism. Not only were the brutal Manchus unable to change the customs of the defeated Chinese; the latter managed to convince the barbarians of the superiority of their own ways. Thus, through rational persuasion, the Chinese demonstrated the superiority of reason over violence. Helvétius in 1774 revived this argument, noting that the Manchus had to adjust to eternal China's Empire when their new dynasty was founded (i.e., the Qing or Manchu dynasty, 1644–1911).

The amusing side of this intense, high-level debate is that none of the Enlightened intellectuals had any direct experience with China. Jesuits had the monopoly of information: Reports by Christian missionaries were almost the sole source of knowledge about that country (with the rare exception of Chinese who moved to France, like the famous intellectual partner of Voltaire, Huang Jialu). This debate is extremely interesting because it provides insight into the contradictory self-consciousness of Europeans, in which understanding China is a relevant matter, even if it was merely one facet of a broader Eurocentric philosophical debate.

Was André Malraux correct in writing that "the Europeans understand of China only what is similar to Europe" (cited in van Staen 2016)? Not exactly, provided that we read carefully the work of Montesquieu, as an advocate of the distinction between the assessment of normative values and knowledge. Montesquieu offers an early example of the comparative sociology of power and political customs in his *Esprit des Lois*: His comparative studies constitute a powerful indirect argument in favor of tolerance and respect for the other, as Raymond Aron (1967: 8) has pointed out.

How does Montesquieu reconcile his normative preference for the English constitutional model and liberty with his own objective approach as a sociologist of political regimes? On the one hand, he is critical of China's extreme social inequalities, tortures inflicted on people convicted of crimes, widespread superstitions, and passivity during the Manchu invasion. To him it seems that the principle of fear is the best explanation for the prevailing despotism in China. On the other hand, Montesquieu underlines China's capacity to maintain traditional customs even in the wake of historical incidents like the invasion. The resilience of Chinese civilization throughout the country's long history provides an example of the incipient autonomy and fledging self-determination which are essential components of modernity. Finally, in spite of his criticism of China, he recognizes that this resilience and stability can be explained only on the basis of the concept of "mixed government" (combining aristocracy, monarchy, and democracy, and wedding the principle of fear with the principles of honor and virtue). To Montesquieu, the inability of civil society to participate freely in public affairs does not appear to have been a serious obstacle to stability.

Here we have moved beyond the relativism of Las Casas in the case of the Indians. Montesquieu addresses the question of the stability or sustainability of political regime-types that differ from constitutional monarchy and Western modernity. Was the way to understanding alternative modernities thereby opened? Yes, it was, as evidenced by the emergence of curiosity about and multidisciplinary knowledge of other continents. However, as argued at the beginning of this essay, knowledge and understanding are independent of both values assessment (is it an inferior or superior civilization?) and the character of the relationship with the other (cooperation or domination?). For example, the conquistador Hernán Cortés de Monroy was interested in acquiring a deep knowledge of Aztec civilization, but only in order to rule the Indians more effectively. Furthermore, ethnology, cultural anthropology, and comparative sociology expand rapidly in the 19th century, notably in Britain (Morgan, Frazer and others) as instruments for making British imperialism more efficient. The key variable is the changing link between knowledge, values assessment, and the evolving nature of the relationship from domination to trade.

Myth and reality of the "universal confraternity of trade"

A relativist approach to knowledge often has characterized the mental world of trade relationships. Is trade part of the common core of all different types of modernity (Meyer, this volume)? International relations based on trade are regulated by the global language of rational choice calculation and reciprocity. Concurrently with the development of cultural and political relativism, European free trade relationships expanded towards East and West. The geographical centers of this European idea of trade as a key factor of international relations were Amsterdam and London, while the theoretical background was provided by Montesquieu, Locke, and Ricardo, notably in his 1817 chef d'oeuvre, *On the Principle of Political Economy and Taxation*. Ricardo argues that industrial specialization strengthens each nation's comparative advantage. Such specialization, combined with free international trade, always creates a win-win situation. The English practice of inserting a most favored nation clause (MFN) in trade agreements implements this approach worldwide. During the 19th century, Richard Cobden, the most famous leader of the political current of economic pacifism, defended free trade as the driving force behind global peace, following Montesquieu's invocation of "*le doux commerce*."

In Europe, liberalizing trade in the framework of the Westphalian inter-state system appeared to be a way out of the destructive, irrational emotions that were manifested in the wars of religion. Inside and outside of Europe, trade entails compromises concerning interests, transactions based on reason and calculation, mediation, and the exclusion of passions and disputes over values. According to Marx, money can build solidarity between enemies. When in 1600 the Japanese Shogun Tokugawa decided to expel the Jesuits and nearly all other Western visitors, only the Dutch traders remained welcome in Japan.

It was Montesquieu who originated the project of establishing a universal confraternity of trade and bringing homogeneity of shared rational-choice-calculation into a world of cultural diversity, yet without undermining the latter. The idea of trade as a unifying but not culturally disruptive set of relationships forms the essential background to classical modernization theory.

However, calculations based on rational choice can offer at best a narrow, partial description of the motivations behind human actions: wars, persecutions, genocides, terrorism, solidarity, love, and hatred very rarely have merely rational choice-type motivations. And free trade quickly reveals itself as something other than fair trade. Furthermore, a few decades after Montesquieu's death, Immanuel Kant penned an open criticism of the combination of free trade and Western colonialism and, in consequence, justified their rejection by non-European peoples (Kant 1795: Third Definitive Article).[1] The negative legacy bequeathed by a Europe pursuing worldwide imperial trade policies contradicts the optimistic idea of trade as a win-win game. Time and again, we witness the revival of an ancient paradox: Internal European stability combined with external hierarchy-building and asymmetric power-consolidation.

The European powers in the 16th century, in spite of having only slight technological and scientific superiority over peoples abroad, notably in Asia, were able to create a twofold Eurocentric world. On the one hand, they maintained a relatively stable internal diplomatic balance of power preventing and diminishing conflicts until 1789, and then once again, after Napoleon, until 1914. We should not undervalue the outcomes of such policies in terms of modern civilization: hard state sovereignty through multilateral cooperation, a common diplomatic culture, and a shared set of institutions. On the other hand, the internal stability of the powers party to the Concert of Europe sparked and abetted the construction of modern overseas Empires. Many scholars (e.g., Pagden 2002) regard those empires as the triumph not only of European technology over the rest but also as the victory of a certain vision of modernity. By 1914, more than 84% of the land on the planet was occupied or controlled by European empires or by the USA (Cuba and the Philippines). This age of Western imperialism (Hobsbawm 1975) lingers on in the memories of the dominated peoples of Africa, Asia, and Latin America as a tragic time of expropriation, exploitation, and cruel enslavement.

From utilitarian specific reciprocity to diffuse reciprocity: The origins of an alternative European way

In the centuries of European modernity, the old continent also has been the crucible of an alternative idea of international relations. Immanuel Kant developed an innovative approach combining the liberal concept of trade with the necessary framework of universal values from legal institutionalization to respect for human rights. In modern terms, this move suggests that the specific reciprocity typical of rational calculations in trade has been upgraded to the level of "diffuse reciprocity" as a condition for creating sustainable, peaceful cooperation among peoples. The shift implies three developments of specific reciprocity[2]: against the immediate demand of return, towards (1) a more extended dimension of time, (2) a complex vision of the social exchange linking multiple issues, and (3) enhanced mutual trust, e.g., trust that one will be repaid in the long run (Keohane 1984, 1986). Diffuse reciprocity is something akin to a cultural norm within a community.

Kant embeds trade relations within the idea of a law of world citizenship, which would guarantee respect for human rights. Is this vision capable of coping with multiple modernities?

Certainly there are many more than the three historical approaches analyzed above: the Roman high-level internal integration combined with a state of war in external relations and shifting European responses to the implications of the first age of globalization, between brutal colonial domination and hierarchical trade (on the one hand) and relativism and an often instrumental understanding of alternative regimes (on the other).

The illusions of a "normative power Europe" and the new challenges

The multilateral institutions (the UN, IMF, World Bank, and GATT) created after the defeat of Nazism, fascism, and Japanese militarism represent a relevant turning point compared with the past. The West, in partial alliance with the USSR and China, seems to have learned some lessons from the great transformation of the early liberal system, at least to a certain extent (Polanyi 1944). Western statesmen realized that the political and economic debacles of the 1930s had something to do with the hard forms of protectionism introduced after 1929 as well as with the weakness of the League of Nations, which virtually collapsed in the interwar years.

Rather than reviving old-fashioned *laissez-faire* liberalism, the new multilateral institutions frame an "embedded capitalism" (Ruggie 1992), in which trade and international economic relations are institutionalized more than ever. The general principle of conduct in this new regime is diffuse reciprocity, as anchored in both the Universal Declaration of Human Rights of 1948 and in chapters VII and VIII of the UN Charter. Both give evidence that there is now a common willingness to integrate trade with various forms of mutual trust to provide a more solid background for deeper, more sustainable multilateral cooperation. Moreover, multilateral institutions form the context in which notions of good governance and multiple modernities gradually are refining the Western liberal model (Meyer, this volume).

The network of multilateral institutions not only supported a variety of capitalist models and domestic Keynesian orders during the golden decades of welfare capitalism (Hall & Taylor 1996). They also expanded them by widening various networks of multilateral cooperation at the global level, as mentioned above, and supporting new forms of global governance, notably after the end of the Cold War and the collapse of communism.

Furthermore, in the context of these multilateral institutions, the idea of consolidating peace through expanded forms of multilateral cooperation at the regional level was put into practice, in particular by six countries of Western Europe (a common market instead of the mere free trade area, as proposed by the United Kingdom). However, Spinelli's idea that the unity of Europe would mark a step towards global peace quickly revealed itself to be more problematic than expected. Certainly, it was unachievable during the Cold War, because of the confrontation between the two global nuclear superpowers.

When did the EC/EU finally become capable of linking the internal peacebuilding community it had created with an external global agenda of peace and cooperation? It was only after 1989–91 that Europe was able to emerge from the inward-looking, low-profile era following the tragedies of World War II. The collapse of the Soviet Union and the partial withdrawal of the United States made it possible for new global actors to emerge (e.g., China, India, and Brazil), including an unprecedented kind of global actor, the European Union.

In this context, the innovative philosophical facts of shared and pooled European sovereignty and of internal diffuse reciprocity are matters of controversy, as is their place and role within a non-European world. How do we define this strange political animal and address the historically unresolved challenge of a true coherence between internal achievements and external relations? On the one hand, the Nye concept of "soft power" is too general, applicable to every international power, including the USA and China. On the other hand, Manners' concept of Normative Power Europe (2002, 2006), combined with the illusion of Europe as a "Post-Modern State" (Cooper 2000) as opposed to a Westphalian, sovereignty-enhancing multipolar planet, appears suspiciously Eurocentric. Therefore, it increasingly has been questioned by non-Europeans (Jain & Pandy 2016, Fitriani 2015, Chen 2013, Acharya 2013b) as well as by European scholars and decision-makers. Such an ambiguous concept, often combined with an assertive, unidirectional promotion of European norms, inevitably has been rejected by the southern part of the world. Its decline, isolation, and paralysis are confirmed by many phenomena, most notably by the most dangerous challenge, the Islamic fundamentalist rejection of Western modernity and of modernizers in the Islamic world itself.

The arrival on the contemporary scene of a religious-political fundamentalism not only denies modern core values altogether, but also supports, along with other factors, extreme forms of holy war and terrorism. For fundamentalists, intercultural dialogue, tolerance, mutual understanding, good governance, and multiple modernities are not basic "virtues" allowing for compromises, but Western values, while Islam offers the true background for a global alternative and revolt against modernity. Some scholars understand it as a symptom of a shift towards a clash of civilizations (Huntington 1996) or towards a new episode in the 2,500-years-long war between West and East (Pagden 2009). But the first targets of ISIS, and in general of the *jihad,* are the victims of the internal Sunni–Shia conflict, modernizing Muslims condemned as apostates, and relatively tolerant Muslim societies threatened by Islamists, like Indonesia, Turkey, and Tunisia. Furthermore, the fundamentalists fail to realize that there are large Islamic communities within Western societies and secularizing tendencies within Islamic countries.

A relevant Christian cultural stream, based on Pope Paul VI's 1964 Encyclical, *Lumen Gentium,* stakes its hopes on an inter-religious "dialogue between all the sons of Abraham," "forgetting the past, for mutual understanding and joint defense of social justice, moral values, peace, and freedom." However, Muslims tend critically to identify modernity and the West with Christianity, and to ignore the distinction between theology and law (Cardini 1994). After many decades of disappointments, and bearing in mind the diversity within the Islamic world, we should not regard religion as the primary cause of conflict. Instead, it is systems of governance and public policies that now look like the crucial problem. For example, ignorance of a then-modern idea of "isonomy" among the Persians (as Herodotus in his *Histories* already defined the concept of political equality in the mid-5th century BCE) may have sharpened the enmity between them and the Greeks. Likewise, some contemporary experts argue (Roy 2016) that the passive acceptance of theocratic tyrannies or façade democracies (Iran, Saudi Arabia) and the rejection by some countries of both the UN Universal Declaration of Human Rights of 1948 and the UN Covenant of 1966 as products of Western imperialism was based on total lack of knowledge about their potential advantages for ordinary people. After all, these documents guarantee, in principle, secularization, abolition of the death penalty and torture, respect for minorities, basic freedoms, and social rights to welfare.

Toward a new and distinctive European path to multiple modernities: Regional good governance and intercultural dialogue?

Under what conditions can Europe revive its distinctive political identity and provide a new, more effective intercultural message? First, a clear differentiation of the contemporary European message of hope and dialogue from the largely disappointing and arrogant "Western modernity" looks like it would be a precondition. The European revival can no longer be a renewal of the old America-centric modern liberal state and society of the boom years after World War II. Second, economists emphasize that Europe – like China and other global actors – should use its economic surplus for investments and expansion rather than for consumption. However, this approach risks producing a narrow understanding of the European role.

According to a voluminous literature, this new role necessarily must involve a European revival within a pluralistic "world of regions," in which several models of capitalism will interact in different ways with the US imperium (Katzenstein 2005). Contrary to Huntington (1996), it does not mean a world of hostile, mutually closed civilizations, but a more pluralistic and multi-level globalized system in which universalism definitely would be a marginal, dying phenomenon if it were identified with the West alone and not combined with various kinds of regional relativism.

Regional relativism also means capitalist diversity. That diversity finds expression not only in the contrast between "coordinated market economies" and "liberal market economies," but also between Rhenish capitalism and the Anglo-Saxon neoliberal model, and – during the recent economic crisis – the ordo-liberal/social democratic model of Central and Northern Europe and the neo-liberal transatlantic wave. This means not only that there are many paths towards the same modernity, but that there are in fact multiple modernities (Meyer, this volume).

However, there is a second fundamental condition for pluralist global governance of an increasingly diverse world, one that is more institutional. Over two decades ago, the Swedish social scientist Göran Therborn concluded his masterwork, *European Modernity and Beyond*, by evaluating Europe's achievements in nation-building, the rule of law, industrialization/wealth-creation, democracy, and the welfare state (1995). But in addition to these observations, he offers an innovative and fairly positive overall assessment of Europe's place in the world. True, he says, the Continent has only one-eighth of the global population; it has been forced back to its original "home" by defeats in its wars of decolonization; it remained dependent on the USA for its nuclear defense until 1991, and found its traditional cultural influence weakened. Yet surprisingly (for a rather euro-skeptical scholar) Therborn emphasizes what he considers the single issue in respect to which Europe again provides the world with a modern and distinctive message: the existence of a supra-nationally institutionalized grouping of neighboring states that has established peace, democracy, and prosperity beyond the Westphalian principles of modernity and the centrality of the sovereign state. Of course, in the future the European Union needs to provide the means for its internal (against terrorism) and international security in a more responsible way. However, Therborn joins a large and diverse cohort of social scientists in underlining this new, potentially universal, European message. That message seems especially paradoxical considering that it emanates from a continent that theoretically (with Machiavelli, Bodin, and Hobbes) and practically invented the hardest forms of national political sovereignty: the nation-state and the Westphalian paradigm.

To understand why and how this innovative idea of a democratic, supra-nationally binding governance at the regional level was able to spread so widely to Latin America (e.g., MERCOSUR), East Asia (ASEAN), and Africa (the African Union), one must mention both the residual European soft power (influencing for example other regional and global regimes like the WTO and the COP 21) and the EU's interregional policies towards other regions (e.g., CELAC, ASEM, and ACP).

By now, it is clear enough to the international research community that it is no longer a matter of asserting a new kind of normative "European model" for the planet. Supra-nationality still built upon national sovereignty has to be adapted further to varied continental conditions. The international literature also suggests the concept of "regionness" (Hettne et al. 1999) to emphasize the profound roots and variety of local cultural and historical sources of regionalism (policies and polities) in the various parts of the contemporary world. Hettne's approach thus constitutes a radical criticism of the idea that regionalization is just a subset of economic globalization. Regionalism means that possibilities for a politics beyond the state remain open. It also suggests that one can build a polity that is capable of coping with regional and global challenges as well as blazing alternative paths to modernity.

Europe can contribute to global governance only by communicating the most innovative features of its distinctive way of theorizing and implementing multilateralism: For the first time, internal and external multilateralism may be consistent and coherent, with internal multilateralism serving as the background for external multilateralism. The Europeans are following the Kantian idea of "federation" (Kant 1795) that should not be confused with Montesquieu's perspective of a federal state in the making – one that very much influenced the US Constitutional Convention of 1787, but is not relevant for the EU. Kant, in his theory of perpetual peace, explicitly opposes a *Staatenbund* to a *Bundesstaat*. The former describes a special kind of confederation of states at the global level, combined with what he calls "cosmopolitanism," a term that actually means emphasizing the crucial role of transnationalism: the dialogue among individuals, a set of people-to-people networks, and cooperation, which collectively go far beyond the mere practice of free trade as suggested by Montesquieu and Ricardo.

The European message is that consolidating peace only will be possible once all the societies of the world are ruled by republican, human rights-respecting, tolerant, and representative governments. That message is no longer an ivory-tower concept derived from a type of cosmopolitanism that underestimates the role of states, the diversity of peoples, and the importance of "good governance." It is a kind of cosmopolitan constitutionalism, based not only on equal interstate relations and binding arrangements, but also on transnational, cross-border ties and networks among citizens belonging to different nations and defending both individual and social rights. It is the way followed by the EC/EU after 1950; however, while during the first decades it operated via a rather inward-looking process of peace- and democracy-consolidation, in the 21st century it can be the core of a new, more understated, form of universalism.

After the end of the Cold War and of the bipolar system based on the mutual nuclear threat, the EU was challenged to show the external dimension of its internally successful multilateralism. This external projection has less and less to do with Westernization, because after the presidency of George W. Bush and now the first months of Donald Trump's term, Europe's foreign policy is taking place against the backdrop of an increasingly "divided West": i.e., a deeper transatlantic rift as far as policies, values, and principles are concerned (Habermas 2006). With Trump, the US is simultaneously questioning both its traditional internal multicultural diversity and its external multilateralism (Ikenberry 2017). Consequently, the EU is challenged to represent the legacy of both internal diversity and external multilateralism (trade, environmental policy, and financial good governance).

However, the EU will respect external diversity and multiple modernities to the extent that it proves itself able to manage internal differences through a deeper brand of multilateral cooperation. For the first time in history, European internal stability and integration are no longer constructed at the price of external hierarchy-building. The relationships with both the near abroad and far abroad might take into account this new opportunity. How could that be done? In contrast to the early, very naive export of the EU regional model, many failures, shortcomings, disappointments, and frustrations have helped give birth to a new approach aimed at fostering a bi-directional dialogue that would be the polar opposite of a clash of civilizations.

Comparative research shows that the European message of secularized and pluralist modernity can be a framework for law-based reconciliation between previous enemies, conflict-prevention, and democratization. But it cannot be exported as a "model"; it can only be suggested as a source of inspiration for other regions in the context of a balanced and equal structure of communication. Regionalism and regional values, perceptions, and identities vary considerably according to different traditions and civilizational contexts. And that offers the background for alternative regionalist ways and various kinds of modernity, characterized by alternative "cognitive priors" (Acharya 2013b). First, as Thomas Meyer rightly points out in the opening chapter, one consequence is that "it is neither informative analytically nor justified normatively to draw one general demarcation line between countries that stick to the Western set of political institutions and the rest of the world." Second, it is increasingly evident that alternative or different cognitive priors do not exclude mutual, respectful interregional communication among countries and regions (Telò et al. 2015), nor do they rule out convergences in the common struggle for building or protecting global public goods (better economic governance, climate change mitigation regimes, offensives against criminality, and terrorism; see Kaul 2016).

But by now some new problems have arisen. Power politics among the big powers has ratcheted up pressure on regional organizations. On the one hand, they are burgeoning due to the quasi-paralysis of global multilateralism (UN, WTO, IMF, and World Bank). On the other hand, regional entities are becoming more ambiguous, following not only alternative, but also competing and mutually exclusive paths (Telò 2016). The comparative regionalist literature converges in recognizing that the process of building regional groupings of neighboring states constitutes, in and of itself, progress towards regional peace and the creation of multicultural entities. Yet it is by no means a guarantee that democracy and sustainable development will come about. "Modern" regional democratic institutions are not exported to other regions from outside. Regional groupings, even if effective in conflict-prevention and economic cooperation, are not always based on truly multilateral rules and procedures, democratic legitimacy, and respect for human rights. Indeed, the very opposite is true; hence, the international reality of regionalism is more variegated than it was during the liberal decade after 1989. The sequencing in the process of institutionalized regional modernization is still controversial: do regional institutions consolidate and efficiently frame the transition to domestic democratization and the national modernization process? Or do national democracies, internal social peace, and reduction of poverty offer the indispensable background for efficient and legitimate regional institutions?

Comparative analyses suggest divergent conclusions when considering recent forms of regional organization such as the Shanghai Cooperation Organization (the SCO), the "Eurasian community," or the Gulf Cooperation Council, which are characterized by an understanding of politics often instrumental to regional powers. New regionalist organizations have drifted away from this top-down dimension of politics, preferring a more democratic, bottom-up form of political integration, a shift that also could play a role in compensating for economic decline. That is happening in ASEAN, MERCOSUR, and the European Union. However, what is needed is a deeper involvement in institutionalized cooperation by the greatest powers. For example, China is involved in various regional entities, including ASEAN+1, ASEAN+3, ASEAN+6, and SCO, and has engaged in trilateral cooperation with Japan and South Korea. However, twenty years of European interregional dialogue with the Chinese elites increasingly has been addressing the crucial challenges of developing a common language regarding respective notions of multilateralism, sovereignty, binding supranational institutions combined with voluntary self-restraint, and evolving balances between individual and social rights (Telò, Ding & Zhang 2017). And we know from experience that a true dialogue always presupposes that the sides previously have established a certain distance from one other (Jullien 2016).

Much to the observer's surprise, thinkers whose backgrounds differ markedly such as Amitav Acharya and John Ikenberry (2017) are asserting the idea of "the end of the liberal order." If their analyses are confirmed, Europe is being challenged to represent in a critical and constructive way the best achievements of its history by combining internal pluralism and external multilateralism. But it is also being encouraged to assert its own idea of modernity within a rapidly changing, more regionalized, and conflictual world.

All in all, the single most original ideas advanced by the Europeans in the second half of the 20th century were those of supra-nationally binding institutions, support for cooperation and integration among neighboring countries, the stabilization of democracy and prosperity, and the consolidation of peace. This bundle of ideas has been exported and emulated in several continents, with the development of new regional organizations during the past 50 years as well as by the supra-national evolution of some global organizations like the WTO and the IPCC. However, those ideas have been implemented only in part. Moreover, they often have been attacked as stagnant by new nationalists who advocate a narrow understanding of the old Westphalian paradigm. Can a pluralist, deeper form of regionalist post-hegemonic multilateralism be not only the successor to Kantian cosmopolitanism but also the bridge-builder for better global governance?

Notes

* ***Editor's note:*** For those unfamiliar with the term, "normative power" refers to the ability of a group to enforce or influence conformity to its behavioral norms.

1 "But to this perfection compare the inhospitable actions of the civilized and especially of the commercial states of our part of the world. The injustice which they show to lands and peoples they visit (which is equivalent to conquering them) is carried by them to terrifying lengths. America, the lands inhabited by the Negro, the Spice Islands, the Cape, etc., were at the time of their discovery considered by these civilized intruders as lands without owners, for they counted the inhabitants as nothing. In East India (Hindustan), under the pretense of establishing economic undertakings, they brought in foreign soldiers and used them to oppress the natives, excited widespread wars among the various states, spread famine, rebellion, perfidy, and the whole litany of evils which afflict mankind.

"China and Japan (Nippon), who have had experience with such guests, have wisely refused them entry, the former permitting their approach to their shores but not their entry, while the latter permit this approach to only one European people, the Dutch, but treat them like prisoners, not allowing them any communication with the inhabitants. The worst of this (or, to speak with the moralist, the best) is that all these outrages profit them nothing, since all these commercial ventures stand on the verge of collapse, and the

Sugar Islands, that place of the most refined and cruel slavery, produces no real revenue except indirectly, only serving a not very praiseworthy purpose of furnishing sailors for war fleets and thus for the conduct of war in Europe. This service is rendered to powers which make a great show of their piety, and, while they drink injustice like water, they regard themselves as the elect in point of orthodoxy.

"Since the narrower or wider community of the peoples of the earth has developed so far that a violation of rights in one place is felt throughout the world, the idea of a law of world citizenship is no high-flown or exaggerated notion. It is a supplement to the unwritten code of the civil and international law, indispensable for the maintenance of the public human rights and hence also of perpetual peace. One cannot flatter oneself into believing one can approach this peace except under the condition outlined here" (Kant 1795, Third Definitive Article).

2 According to Keohane (1989; cited in Kramer 2014), specific reciprocity refers to "situations in which specified partners exchange items of equivalent value in a strictly delimited sequence. If any obligations exist, they are clearly specified in terms of rights and duties of particular actors."

8

THE ROLE OF THE UNITED NATIONS' ALLIANCE OF CIVILIZATIONS IN BUILDING CULTURALLY INCLUSIVE SOCIETIES IN THE 21ST CENTURY

Nassir Abdulaziz Al-Nasser

Not everyone is familiar with the Alliance of Civilizations and the role it plays within the United Nations. By way of introduction, I would like to highlight the United Nations Charter, which is the effective Constitution of the United Nations and the framework for all its component programs, institutions, and entities.

The United Nations was founded on the ashes of the destruction wrought throughout the planet by the Second World War, a war that killed some estimated 80 million people, the majority being civilians. The Charter stresses the faith of the people of the United Nations in human rights, the dignity and worth of every human being, and the equal rights of men and women. It stresses the need for international co-operation in solving international problems of an economic, social, cultural, or humanitarian character. The Charter underlines the importance of tolerance and of promoting and encouraging respect for fundamental freedoms for all, without distinction as to race, sex, language, or religion.

Since its creation, the United Nations has pursued these goals relentlessly, and we all know that this will remain an uphill battle. Reaching the top of the mountain requires unceasing efforts and contributions from all participants. Governments, non-governmental organizations, academia, religious leaders and private businesses all have to join forces and coordinate their efforts so that peace and security can be guaranteed for all.

This is why the initiative of the Institute of European Studies of Macau to hold a conference that covers the significance of multiple modernities and the need to expand good governance and human security in our multicultural world is so important. This initiative fully supports the goals of the United Nations and the vision of the United Nations that humanity can flourish in a peaceful and secure world only if we pool all our efforts.

How different is the world that saw the birth of the United Nations from our world of today! Our world now is marked by the impact of globalization and instantaneous communication throughout the world. Another noticeable change is that there are few – or no longer any – barriers between states. The two wars that decimated millions of people in the 20th century were conflicts between nations. Today, we can see that threats to human security are rarely confined by borders and rarely involve one state against others. The world has been transformed beyond anything our founders might have imagined. In that context, one should realize that the Alliance of Civilizations was created more than ten years ago to challenge Huntington's prophecy about the "clash of civilizations" at a time when the world was already polarized. This polarization stemmed from injustice and inequality which, in turn, led to violence and conflict that threatened international peace. Hence, UNAOC was created to build bridges between societies in order to promote dialogue and understanding and forge the political will to address the world's imbalances.

This new millennium started with tremendous optimism. With the end of the Cold War, the tension between the East and West had eased. Global cooperation had reached a high-water mark. The leaders of the world met in New York and adopted the UN Millennium Declaration. They affirmed their shared values and committed their nations to a set of goals for development and for the eradication of extreme poverty. They affirmed their commitment to disarmament, security, human rights, and the protection of the environment. They agreed to a series of time-limited targets, with a deadline of 2015, which have come to be known as the Millennium Development Goals (MDG). By 2015, the world had made much progress in achieving these goals, but the results have been uneven. In 2015, participating countries reviewed the MDG achievements and decided to build on the MDG and expand them further.

Through a deliberative process that included member states, academia, and civil society, the United Nations adopted the Sustainable Development Goals (SDGs), with a target date of 2030. The proposal contained 17 goals with 169 targets and 230 indicators, covering a broad range of sustainable development issues. These included ending poverty and hunger, improving health and education, making cities more sustainable, combating climate change, and protecting oceans and forests. The last goal of the SDG, goal 17, highlights the importance of partnerships for insuring that all the goals will be achieved.

At the same time that the United Nations was committing member states to a program of improving living conditions for all, chaos was shaking regions of the world. The beginning of the third millennium was marked by the horror of September 11. Terrorist groups determined to inflict death and turmoil started to emerge in Europe, Africa, and the Middle East. Their gruesome attacks make no distinctions among races or religions. Extremist groups like ISIS, Al Qaeda, or Boko Haram are not confined to a specific region. They recruit adherents through the Internet, social media, and all the new forms of communications that have emerged over the last twenty years.

We are still trying to understand why young people are attracted by the message of hatred of extremist groups and why they are ready to sacrifice their lives while committing horrific acts of destruction. Violent extremists have been able to recruit over 30,000 foreign terrorist fighters from over 100 member states to travel to the Syrian Arab Republic and Iraq, as well as to Afghanistan, Libya and Yemen. The Internet brings people from across the world together for the better but also for the worse.

Civil wars and the destruction wrought by extremist groups have led to a refugee and migrant crisis on a scale not seen since World War II. Millions of people have fled the territory controlled by terrorist and violent extremist groups. Hundreds of thousands are forced to leave their homes and their villages where generations of their forebears lived in peace and harmony. Recently, 6,000 people were rescued in one day as their boats capsized crossing the Mediterranean, but in 2016 more than 3,500 drowned! These huge refugee movements have led to the resurgence of racism, intolerance, and xenophobia in the communities where refugees try to settle or simply transit through. You see them desperately trying to protect their families and taking enormous risks in the hope of a better life. Whether in the bombing of Syria or in the crossing of the Mediterranean, children are dying every day.

We know that the new millennium was accompanied by the growth of fear and misunderstanding between Islamic and Western societies. The heightened instability of coexistence between these groups of people with divergent backgrounds has led to exploitation by extremists throughout the world, of which the most destructive expressions are violent acts of terrorism. But we all believe that the majority of the world rejects extremism and embraces diversity. What could the United Nations do to support inclusiveness and counter the animosities that pit people from different religions and cultures against each other?

It is against this dramatic background – one that seemed to pit the West against the Muslim world – that the Alliance was born. The Alliance was established in 2005 as the political initiative of Mr. Kofi Annan, former UN Secretary General. His initiative was co-sponsored by the governments of Spain and Turkey. Mr. Annan formed a high-level group of experts to explore the roots of polarization between societies, cultures, and religions.

UNAOC was designed to be the arm of the United Nations that would assist in diminishing hostility and promoting harmony among the nations, especially between the West and the Muslim world. The Alliance would call for international action against extremism through the forging of international, intercultural, and interreligious dialogue and cooperation. The objective of the Alliance would be to promote understanding between nations and groups of different cultures and religions on the assumption that building bridges and holding dialogues are the best means to prevent conflict and promote inclusive societies. The Alliance can serve as a mediator in many conflicts in which differences of culture or religion are involved.

Currently, 145 countries belong to the Group of Friends of the Alliance. Group members concluded that policy recommendations were not sufficient. Their view was that tensions across cultures have spread beyond the political level into the hearts and minds of populations. To counter this trend, the Group recommended that the Alliance should focus its work on four specific areas in which advocacy and targeted projects could make a difference: education, youth, migration, and media.

A quick explanation as to why the Alliance is focusing its work on these four areas follows.

Education is the key to opening the minds of young people and helping them to understand that living and working with people of different cultural, ethnic, or religious backgrounds enhances the richness of life. It allows for a flow of creativity and encourages young people to be tolerant with each other. Education helps young people to understand that they face similar challenges and that they need to work together.

Youth is seen as the group most vulnerable to the enticements of extremist groups, which target and recruit new members though false messages of religious extremism. The lack of employment and prospects can leave young men and women alienated and prey to extremism. Eighty percent of the world's young people live in developing countries, and the largest number of these live in the two areas most touched by wars and terrorism: Africa and the Middle East. Children and women are the most affected groups. They suffer from sexual violence and enslavement. In fact, sometimes they are even forced to take part in horrendous acts of terrorism.

Xenophobic, racist, and religiously-biased language in the media can propagate messages of hatred that have particularly negative impacts on immigrant and refugee communities. Hatred-filled media coverage can make it extremely difficult for migrants to be integrated into the societies that have given them refuge. Media can and should provide unbiased information and help to propagate positive messages about diversity.

Migration and refugees have taken center stage at the United Nations. Over the last few years, we have seen how migrants can be perceived as threats to communities that believe that their cultural identity is being shattered by the arrival of migrants and that the latter may limit their job opportunities.

To reach its objectives, the Alliance has mobilized governments as well as non-governmental organizations, grassroots organizations, religious leaders, community leaders, and especially youth leaders. Our common goal is to develop inclusive societies in which diversity is experienced as an asset.

Since its inception, the Alliance has generated growing support from the membership of the United Nations and from international and regional organizations. The UNAOC brings together developed countries of the North and developing countries of the so-called global South, the East, and the West. The Alliance, along with its affiliated Group of Friends, is now the lead entity within the United Nations in assisting countries to address global challenges with interfaith and intercultural dimensions.

Building on the work of my predecessors and since taking the helm of the Alliance in February, 2013, I have developed a vision for the Alliance and set a series of clear priorities. The first priority remains to build on previous achievements and to move forward to accomplish additional goals in the core areas of education, youth, migration, and media. I have put great emphasis on strengthening partnerships and cooperation within the United Nations system as well as with civil society, religious leaders, and especially the media and the private sector. I have also added sports, music and entertainment, and other forms of collective expressions of human values to the UNAOC toolbox.

The Alliance has also developed a framework to support the Sustainable Development Goals, as it is clear that development cannot happen where extremism reigns. Inequalities are not only a major development problem, but also pose a threat to global security. Frustration, marginalization, and exclusion stemming from social inequities and lack of opportunities will impede the achievement of peace and security. The goals of the new Sustainable Development Agenda 2030 demand the participation and inclusion of all people and especially young men and women. The goals can be attained only when cultural diversity and religious differences come to be seen as an advantage and not as a threat. Successful sustainable development can happen only when dialogue replaces conflicts.

Terrorism and violent extremism are a global threat to international peace and security. The Alliance has forged a strong relationship with the United Nations Counter Terrorism Implementation Task Force (CTITF), established by the United Nations Security Council back in 2001. We complement each other's efforts in fighting the horrors inflicted by terrorist groups. The misleading use of religious ideologies is harming the global balance of harmony among nations, cultures, and people. Much of the work of the Alliance focuses on creating alternatives to radicalization and counteracting the messages of extremist organizations that falsify the core of religious beliefs. I am happy to announce that I was selected as a member of the Secretary General's high-level PVE (Prevention of Violent Extremism) Action Group. The selection was the fruit of months of deliberate engagement with the CTITF Inter-Agency PVE Working Group following the instructions of the High Representative that UNAOC should assert and strengthen its presence and contribution to PVE.

My vision is supported by a series of projects. Here are a few examples of how concrete and targeted our initiatives are. Recently in Istanbul, together with The Cooperation Council of Turkic Speaking States (Turkic Council) we hosted an **International Conference** on The Role of Youth in Preventing and Countering Violent Extremism and on Holistic Approaches from Education to De-radicalization. We held panel discussions on the contributions of young people to peace-building efforts and the achievement of the Sustainable Development Goals. We presented concrete case-studies on de-radicalization strategies and programs through the use of innovative methods including social media and sports.

We also have a **Fellowship Program** that aims at fostering intercultural exchange and interfaith understanding by engaging with emerging leaders and young professionals from Europe, North America, the Middle East, and North Africa. The program sends participants from each geographic area to one of the other geographic areas while interacting with a wide range of local actors and partners in every country visited.

The program offers the participants an opportunity to be exposed to a different culture. The participants are provided with the opportunity to understand the plurality and complexity of their surroundings and to understand the host country's culture, society, religion, and media. The participants are exposed to new ideas, and that allows them to challenge their perceptions and to deconstruct stereotypes.

Our **Youth Solidarity Fund** (YSF) supports youth and organizations that foster peaceful and inclusive societies by providing direct funding to outstanding projects promoting intercultural and interfaith dialogue. The funded projects are youth-led and youth-focused but have an impact on the entire communities, often involving religious or political leaders, policymakers, educational institutions, and media organizations.

The projects funded by the YSF target young people from various backgrounds: students, marginalized youth, minorities, youth in rural or urban areas, youth in conflict or post-conflict situations, artists, and activists. The youth-led organizations employ creative methodologies to break stereotypes, improve intercultural relations, and promote a culture of peace. As of 2015, a total of 43 projects had been completed reaching a total of over 800,000 persons in more than 30 countries.

Our **Summer School** brings together youth from around the world to address pressing global challenges within the context of cultural and religious diversity. This educational experience for young civil society leaders is designed to strengthen their knowledge and skills, empowering them to be involved more effectively in building peaceful societies and collaborating across differences, through an augmented network spanning the globe.

During one week, a group of 75–100 participants aged 18–35 engage in workshops, roundtables, and collaborative work focused on fostering diversity and global citizenship; reducing stereotypes and identity-based tensions; and promoting intercultural harmony and social justice. We have operated six Summer Schools so far, offering nearly 500 youth from 119 countries the opportunity to increase their understanding of other cultures and faiths, reminding them of the similarities that unite us despite our differences, and encouraging them to bring about positive social change.

One of our latest initiatives is **The Young Peace-Builders in West Africa**, designed to engage young women and young men from West Africa in an intercultural dialogue and peace-building experience. The aim is to support the growth of networks of young peace builders who are equipped with the tools to address stereotypes, prejudice, and polarization in order to build more inclusive and peaceful societies in their communities and globally.

Another recent initiative is our **Plural+** program, a youth-produced video festival that encourages young people to explore migration, diversity, and social inclusion, and to share their creative vision with the world. Plural+ involves cooperation between the United Nations Alliance of Civilizations and the International Organization for Migration, with a network of over 50 partner organizations that support the creative efforts of young people and distribute their videos worldwide.

In view of the importance of the media in forging positive images of inclusiveness, we have developed key projects in that area. Evaluating information sources requires critical thinking. In collaboration with UNESCO, children and youth from industrialized societies spend twice as much time immersed in electronic media as they do receiving formal education in schools. The impact of the media on forging ethical and social values is huge. We want to emphasize the need to develop strategies for media literacy education that will provide young people with the tools necessary to have a critical view of the information provided by media. We want to help teachers develop curricula that help young people to understand better the messages to which they are constantly exposed.

The media also play a crucial role in the public's perceptions of migrant and refugee populations. Negative portrayals of immigrants and refugees are often found in the media, thus negatively impacting people's views of these communities. We recently launched a **#SpreadNoHate** initiative on how best to counter hate speech through global forum discussions and a social media campaign.

In September of 2016, on the margins of the General Assembly, we held a side event on combating xenophobic and hate language in the media. Xenophobic language disseminated by the media can propagate negative images of migrants and refugees. On the other hand, the media can also highlight the enrichment that refugees and migrants can bring to established communities. The problem is especially acute in Europe, which has seen hundreds of thousands of refugees trying to reach its shores and cross its borders.

As part of its mandate, UNAOC works to combat racist, xenophobic, and prejudiced messages in the media. Since its creation, UNAOC has focused its attention on the role of the media and its impact on minority communities. The media constitute a fundamental force that shapes the lives of individuals and impacts the fate of peoples and nations.

This side event was part of the UNAOC #SpreadNoHate initiative, which is a series of international forums about hate speech and the best practices to counter and prevent it from being disseminated by the media. Our next #SpreadNoHate symposium is planned for early next year in Brussels. I am very pleased to announce that the European Union has agreed to be our partner and sponsor in hosting the symposium.

The problem of migrants and refugees has never been as acute as in recent years. We strongly believe that migration can play a positive role in fostering sustainable development and contributing to sustainable and inclusive societies. But it is true that migration has presented many societies with major policy dilemmas. Many countries of the world are becoming more multi-cultural, multi-ethnic, multi-religious, and multi-lingual. Too often, the local community perceives migrants as a threat. They fear that their cultural heritage no longer will be respected instead of seeing migrants as an asset and an added value. New migrants are seen as an economic drain and a strain on public benefits. They are perceived as unable to adapt to customs and life in receiving societies. Finally, some of them are associated with fears of terrorist attacks. The projects mentioned above are just a few examples of the projects that UNAOC has launched over the last few years.

More than ever, we understand that inclusive societies are the bedrock of a world at peace. Each and every one of the 17 new Sustainable Development Goals calls for dialogue across civilizations and religions. These goals can be achieved only if people, communities, and nations work together across cultures, religions, and ethnic groups.

The Alliance, as a soft diplomacy tool employing advocacy, projects, and its mediation role, is determined to advance the importance of dialogue across civilizations in order to build inclusive societies.

In conclusion, let us once again emphasize that a lot has been achieved in the last two decades thanks to the efforts initiated by the United Nations. However, there is still a lot of work to be accomplished in the 21st century in the areas discussed in this conference, which are poverty reduction, economic development, gender equality and female empowerment, fostering the development of civil societies, promoting fundamental human rights, sustainable development, and international security, combating terrorism and international organized crime, and preventing interstate conflict. In order to build inclusive societies in the 21st century, as well as a peaceful and harmonious world, tangible improvements are necessary through wide-ranging, ongoing, long-term, persistent cooperation of the international community.

9

HARMONY AND HUMAN DIGNITY

A Confucian perspective

Julia Tao[*]

The chapter has five parts:

(I) The concept of ho (和) or harmony in the Chinese tradition
(II) Ho (和) or harmony as a social and political ideal from the Confucian perspective
(III) Mencius's notions of natural equality and natural nobility as basis of human dignity
(IV) Difference and relatedness in Confucian universalistic human dignity
(V) Why harmony requires respect for human dignity

I. *Ho* (和) or harmony in the Chinese tradition

Harmony is an important concept in Confucian thinking. Although harmony does not have a central place in the West, the ancient Greek philosopher Plato (1961) had developed a theory of harmony based on his notion of ideal forms. Unfortunately, given its significance, harmony is arguably much neglected and the most understudied concept in politics and governance, both in the East and the West.

The Chinese word for harmony is written today as 和 (*he*). To provide a lucid exposition of the rich meaning and profound significance of this concept in the long history of the Chinese culture, this essay will draw extensively on the analysis presented by Yu Kam Por in his seminal article, "The Confucian Conception of Harmony" (2010), which is one of the most rigorously researched and original works on the subject. Yu is an erudite scholar with solid training in both the Chinese classics and Western philosophy. In his article, Yu drew on rich evidence from ancient Chinese texts to substantiate his philosophical analysis of the concept *he* and explore with keen insight its far-reaching significance as a social and political ideal in the Confucian tradition.

According to Yu, ancient writers in China often use analogies to illustrate an unfamiliar concept instead of giving definitions. In his exposition, he cited in detail from the Chinese classics to provide a robust discussion of how three analogies were commonly used by the ancient Chinese to bring out the essential features of the concept of harmony as an ethical ideal in the mainstream culture (2010: 18–20).

Cooking is the first of these analogies. The following conversation was recorded in the *Zhouchuan* 左傳 (Zuo's commentary to the *Spring and Autumn Annals*) in 525 BCE between the Duke of the state of Qi 齊 and his prime minister Yanzi 晏子 regarding what constitutes a harmonious relation between a ruler and his minister. The exchanges highlight the important distinction between harmony and complete agreement.

> The Duke was complaining that among his ministers, only Ju of Liangqiu 梁丘據 was harmonious with him.
>
> The duke said, "Only Ju is in harmony (*he* 和) with me."
>
> Yanzi replied, "Ju is in complete agreement (*tong* 同) with you. How can he be in harmony with you?"
>
> The duke said, "Are harmony and complete agreement different?"
>
> Yanzi replied, "Different indeed! Harmony is like making soup. Water, fire, vinegar, minced meat, salt, and plum are used to cook the fish and meat. These are heated using firewood and brought into harmony by the chef, who uses different flavors to achieve *a balance*, providing what is deficient and releasing what is excessive."
>
> *(Year 20 of the "Duke of Zhao昭公" in Yang 1981: 1419–20; cf. Legge 1960, vol. 5: 684)*

Music is the second common analogy used by the ancients. Yanzi used the music analogy in the same conversation recorded in *Zhouchuan* to emphasize that harmony is essentially based on diversity. He pointed out that a piece of music is made up of different and contrasting elements. There are different instruments, volumes, melodies, pitches, and voices. Although these elements contrast, they also complement one another, as Yanzi explained:

> The former kings balanced the five flavors and the five notes in order to calm and settle their hearts and minds and perfect their government.
>
> Notes are like flavors. [In music] there is [human] breath, the two types of dances, the three classes of songs, the materials [to construct instruments] gathered from four corners, the five-note scale, the six-pitch pipes, the seven-note scale, the winds of the eight directions, and the nine songs of praise – all of which *perfect* one another.
>
> [Among notes] there are clear and broad, small and large, short and long, agitated and sedate, sad and joyful, strong and yielding, slow and fast, high and

low, initiation and conclusion, and proximate and distant – all of which balance one another.

Health is the third common analogy. As Yu pointed out, there is a strong belief among the Chinese that, much like harmony in cooking and in music, harmony of the body depends on the right balance of opposite elements. This explains why, in Chinese medicine, we speak of an unhealthy body as a body breaching harmony (*weiho* 違和). Harmony on this understanding does not imply the elimination of opposites; it actually requires the presence and the appropriate balance of opposites.

Hence, to maintain harmony of the body, there should be, for example, balance between the "hot *qi*" (熱氣) and "cold *qi*" (寒氣) and between the "wet *qi*" (濕氣) and "dry *qi*" (燥氣). The loss of balance will result in disharmony and ill health.

Like harmony in cooking and harmony in music, harmony of the body requires the presence of *different* and often *opposite* elements necessary for one's health and the right balance of different ways of life.

According to Yu, the leading Confucian scholar of the Han dynasty, Dong Zhongshu董仲舒 (ca. 195–105 B.C.E.), went so far as to say that harmony is a principle to govern the world as well as a principle to take care of one's body: "Those who can manage the world with harmony have illustrious virtue. Those who can nurture their bodies with harmony have extreme longevity" (*Chunqiu fanlu*春秋繁露 [Luxuriant Gems of the Spring and Autumn], Book 16; Yan 2003: 292).

Even today, it remains a strongly held belief among contemporary Chinese that the maintaining of balance will result in harmony of the body, which is called health.

Main features of the Confucian concept of harmony

Harmony presupposes difference and diversity

It is clear from these three analogies analyzed by Yu that it is not possible to speak of harmony in the absence of difference or when there is just one single item or thing. Inherent in the Chinese concept of harmony is the notion of difference and diversity.

Harmony in the Chinese account presupposes the existence of different things and different actions, and implies a certain favorable *relationship* among them, which Si Bo (史伯), the grand historiographer of the late Western Zhou (774 BCE), explained to the Duke Huan of Zhen (鄭桓公) as follows.

> When there is monotony of sound, there is no music.
> When there is monotony of things, there is no pattern.
> When there is monotony of taste, there is no delicacy.
> When there is monotony of things, there is no harmony.
>
> *("*Zhengyu*"* 鄭語*, ch. 16,* Guoyu*; Wei 1978: 515–16)*

Complete agreement or sameness does not constitute harmony, as pointed out by Yanzi when explaining why Ju's complete agreement with the Duke of Yi was not a case of genuine harmony:

> What Ju is doing is nothing like this. What you find acceptable, Ju also says to be acceptable. What you find unacceptable, Ju also says to be unacceptable. This is like adding more water to water. Who can eat that kind of food?
>
> It is like the monotonic sound made by musical instruments – who can listen to that kind of music? This is why complete agreement is unacceptable.

Harmony requires balance but does not mean sameness or complete agreement. Sameness, whether in the form of making music with the same type of instrument or demanding the people's absolute agreement to the ruler, is contrary to harmony.

Yu pointed out that in the sphere of governance, Confucius himself also put considerable emphasis on balancing. As he expressed it, "Leniency complements strictness. Strictness complements leniency. Hence, there is harmony in governance" ("*Zuozhuan*" 左傳, Year 20 of the Duke of Zhao; Yang 1981: 1421; cf. Legge 1960, vol. 5: 684).

Heterogeneity is at the heart of harmony

According to the ancient Chinese thinking, harmony is based on heterogeneity. It gives rise to a desirable quality that is not contained in any single element but is nevertheless constituted by its elements and cannot exist without them.

So, for example, in order to make a dish, there must be different ingredients. Each ingredient has its own taste, but different ingredients put together can form a new taste. If we use the same thing to complement the same thing, nothing is accomplished. The right balance of different ingredients is crucial. Simply adding more of the same kind cannot create a new and better taste.

When the different constituents each have their own place, when they interact but do not transgress on one another's realm, there is harmony. Just as it takes different constituents to make a piece of music, it also takes heterogeneity in a society to constitute a harmonious society.

In his conversation with the Duke Huan of Zhen about the fate of the Western Zhou, Shi Bo predicted that the decline of the Western Zhou was inevitable because the government pursued homogeneity instead of harmony by seeking complete agreement from its ministers. In Shi Bo's diagnosis, harmony brings a state of creative balance, whereas homogeneity brings stagnation. He cited many examples to illustrate how heterogeneity is necessary to achieve sophisticated and desirable outcomes.

> So the former kings sought their queens from a different clan, looked for resources in faraway places, selected ministers who would remonstrate with

them, and explained things with a broad range of examples. This is because they aimed at *harmonious agreement* (*he tong* 和同).

(“Zhengyu” 鄭語, *ch. 16,* Guoyu; *Wei 1978: 515–16)*

As Yu observed, these ancient Chinese accounts of harmony are highly definitive, formative, and authoritative in shaping the Confucian or even the Chinese conception of harmony over the last two and a half millennia. A sharp distinction is made between harmony and uniformity. Harmony is regarded as the source of creativity and transformation, while uniformity is regarded as the cause of stagnation. Harmony is conceived as a dynamic and generative process based on the incessant and organic interplay of different forces, which Yanzi explained and was recorded in this way in the *Guo yu*:

He (harmony) gives rise to new things;
Tung (uniformity) will lead to stagnation.
To balance one thing with another is called "*he*" which will lead to enrichment;
To add to the same thing yet more of the same will ruin the whole.

(“Zhengyu” 鄭 語, *ch. 16,* Guoyu; *Wei 1978: 515–16)*

II. Confucian harmony as a social and political ideal

Following the early Chinese thinkers, Confucians also conceive of harmony as an active and dynamic process in which heterogeneous elements are brought into a mutually balancing and complementary relationship to form new things, orders, or patterns. Harmony does not mean complying with a pre-given perfect order of the world. Instead, it refers to a ceaseless and endless process of constant self-renewal and transformation.

In Plato's *Republic*, perfect harmony only exists in the realm of Forms. Order in the world is achieved and maintained by conformity to the Forms in the transcendent realm. In Confucian thinking, by contrast, there is no such fixed order from Heaven or God, no grand scheme of things to which humanity must conform.

Such a conception of harmony is based on the Chinese understanding of nature as an organic process, a spontaneously self-generating life force (Tao 2005: 71). Nature as a life force is dynamic and ceaseless, consistent and forever changing, transforming and unfolding new contours, new forms, and new lives. We can witness this fact in the endless birth of new things and the ongoing evolution of all things. Human life, also a process of cultivation and transformation, emulates nature.

Thus, harmony as the basis of prosperity is also found in both the natural and the human worlds. It is only by maintaining and preserving harmony and diversity among the myriad thousand things in nature that the natural world can flourish and perpetuate its growth and development.

The Confucian conception of harmony emphasizes the analogy between the importance of maintaining harmony in the natural order and in the social and political order. In the Confucian view, the world is not there just for one item or one kind of thing. It is there for the "myriad things" (萬物). Nothing in the world can claim absolute superiority to all others. Harmony is neither conformity nor submission. As emphasized above, there is no pre-given, fixed world order to which humanity should conform.

The Confucian view reminds us that harmony as a social and political ideal cannot be created by an order imposed top-down or by military might or coercive control. It can only be achieved by respecting the diversity in the world and addressing the legitimate claim of each party to ensure optimal space to flourish.

Diversity, different perspectives, different values, different opinions form the background or platform of harmony. The Song dynasty Confucian scholar, Cheng Yi 程頤 (1033–1107), explained it in this way:

> Everything in the world has its due. If it gets its due, it is in peace. If it loses its due, it causes trouble.
>
> The sages were able to make the world smooth and peaceful not because they made rules for everything, but because they succeeded in *letting everything get its due*.
>
> *(Cheng & Cheng 1981: 968)*

Only by accommodating others and learning from those who are different can one really enrich and enlarge oneself and thereby prosper. It is clear that from the Confucian perspective, harmony or *he* does not mean the suppression of individuality or the attainment of a steady state.

To enable everything in the world to get its due and to flourish in harmony in togetherness with one another, justice (義) is also a necessary component of harmony. Xunzi 荀子, another ancient Confucian scholar, argued that "assignment of parts on the basis of justice brings harmony" (以義分則和). He elucidated this point as follows:

> *How can people live together?*
> Answer: assigning parts (分).
> *Why does part assignment work?*
> Answer: *Justice* (義).
> *Assignment according to justice constitutes harmony.*
> *When there is harmony there is togetherness.*
> *When there is togetherness there is concerted effort.*
> *When there is concerted effort there is strength.*
> *Where there is strength there is successful accomplishment.*
>
> *("Institutions of the King" ['Wangzhi王制] in the* Xunzi; *Liang 1974: 109)*

Moreover, harmony also must be distinguished from "sameness" or "conformity." In *The Analects* (論語), Confucius himself drew a clear distinction between a moral person and a mean one (Lau 1979a 13: 23): "The moral person *ho* (harmonizes) but not *tung* (conforms)/The mean person *tung* (conforms) but not *ho* (harmonizes.)"

Pursuing harmony does not mean seeking uniformity or conformity. The goal of uniformity implies accepting only those who are agreeable to oneself, and excluding those who are not agreeable. This will have the effect of limiting oneself; society and government will come to a standstill and become stagnant because of lack of diversity and difference.

Pursuing harmony implies accepting those who have different values and views from oneself, those who are ready to speak up in a different voice and to disagree in a forthright manner. It is only by accommodating others and learning from those who are different that one can enrich and enlarge oneself and prosper.

Pursuing harmony also does not mean following the crowd or the popular opinion. Confucius has also distinguished the concept of harmony from the concept of *liu* 流 (flow). *Liu* means following the crowd or fashion. The gentleman looks for harmony, but it does not mean that he just follows the majority or fashionable view. As Confucius said: "A gentleman looks for harmony, but he does not go with the flow. Strong indeed is he" (*Zhongyung* 10).

Harmony is not unprincipled compromise. To harmonize is not the same as to bargain through horse-trading or to surrender one's principles. The *jungzi* (君子 moral person) stands his ground and is not moved by popular opinions, much less being shaped by them.

Clearly, Chinese Confucian conception of harmony as a social and political ideal cannot be achieved by eliminating diversity or ignoring justice. Neither can it be achieved by suppressing individuality or denying difference. As we have seen, harmony is a dynamic and generative process based on the incessant and organic interplay of different forces: forever changing, transforming and unfolding new contours, new forms, and new things in the world.

The Confucian understanding of harmony enables us to value difference and appreciate diversity. But harmony cannot be achieved merely on the basis of difference and diversity alone. Harmony as a social and political ideal also requires the ability to recognize our human relatedness and appreciate our moral equality as part of a shared humanity. From the Confucian perspective, harmony must be built on the basis of respect for universal human dignity, our common morality.

III. Human dignity: Mencius's natural equality and natural nobility

In the Confucian moral tradition, we find no equivalent to the concept of human dignity as it has come to be defined in Western philosophical traditions and used in contemporary modern liberal democracies to underpin a system of universal rights. There is, however, strong textual evidence to suggest the existence of a distinct notion of natural moral equality in Mencius's theory of human nature more than 2,000 years ago. In his theory, Mencius孟子 posited a strong notion of "natural nobility" that is intrinsic to all humans and sharply distinguished from "aristocratic nobility" based on social role and status.

According to Mencius, human beings are born with certain common natural inclinations, sentiments, emotions, desires as well as ethical aspirations, etc., which constitute the "*qing*" (情) of the human person. In the Confucian discourse, *qing*, or the natural or real situation of the human person, is referred to as the human *qing* or *ren qing* or the *qing* that is commonly shared by all people (人情) (Yu & Tao 2012).

The Chinese concept *qing* (情) as it is used today is commonly translated as "emotion." In the Confucian usage, it refers to the genuine inner state of one's heart-mind (there being no distinction between heart and mind in the Confucian perspective). These human *qing* refer to our inborn human nature. The concept includes emotions, but is not confined to them alone. In one of the Chinese Classics, *The Book of Rites* 禮記, for example, the following was recorded: "What is human *qing*? Delight, anger, sadness, fear, preference, disgust, and desire. All humans know these seven things without having to learn them" (Liji [The Book of Rites], Liyun).

On Confucian understanding, human *qing* is not merely affective, but also has a cognitive component. Because of these common *ren qing* or human *qing*, we are able to have some minimal knowledge of other people without having to know them personally.

Mencius further argued that what is the most important of these emotions and inclinations that all humans are born with are "the four sprouts of morality": the hearts-minds of **compassion, shame, modesty**, and **right and wrong**. They are all regarded as part of human nature. Beings who do not possess these four moral sprouts cannot be human, as Mencius wrote:

> Whoever is devoid of the heart of compassion is not human,
> whoever is devoid of the heart of shame is not human,
> whoever is devoid of the heart of modesty is not human, and
> whoever is devoid of the heart of right and wrong is not human…
> Man has these four sprouts just as he has four limbs.
>
> *(2A6, Lau 1979b)*

He went on to explain that the heart of compassion is the seed of **humaneness** (仁), the heart of shame is the seed of **rightness** (義), the heart of modesty is the seed of **propriety** (禮), and the heart of right and wrong is the seed of **wisdom** (智). Mencius considered compassion as the most important trait because it is the beginning of the seed or virtue of humaneness or humanity, *ren* (仁). To prove that all humans possess this moral sprout, he invited us to participate in the following thought experiment.

> My reason for saying that no man is devoid of a heart sensitive to the sufferings of others is this. Suppose a man were, all of a sudden, to see a young child on the verge of falling into a well. He would certainly be moved to compassion, not because he wished to get in good graces of the parents, nor because he

> wished to win the praise of his fellow villagers or friends, nor yet because he disliked the cry of the child.... The heart of compassion is the beginning of humaneness.
>
> *(2A6, Lau 1979b)*

In Mencius's theory of human nature, these four sprouts are our moral potentials. They are the essential raw materials from which human morality is to be crafted. In Confucian morality, family love or *qin qing* (親情) is a natural starting point, because family love or *qin qing* between parent and child expresses the most natural form of *qing* of human beings. It is uncalculated and not imposed by rules. It exemplifies the most valuable of human *qing* because of its genuineness, intimacy, constancy, and naturalness. Whereas in Christianity God is conceived as the source of love, in Confucian morality, love roots in, or originates from, "family" or "family life."

The two tools that make love possible in Mencius's moral theory are "extension" and "self-cultivation." That is why Mencius persistently urged people to "treat with respect the elders in your own family, and then extend that respect to include the elders in other families. Treat with tenderness the young in your own family, and then extend that tenderness to include the young in other families" (1A7: Lau 1979b).

The more a person conducts his self-cultivation, the farther his extension can reach and the more lasting the extension can become. Thus, although humanness or compassion begins in the family and in family relationships, its final destination is the general other.

As Mencius explained, the four sprouts of compassion, shame, modesty, and right and wrong are the "four beginnings" or "possibilities" of the virtues of humaneness, rightness, propriety, and wisdom. They are the foundation of human morality. The distinction based on these innate moral tendencies and emotions draws a definitive line between humans and non-humans:

> Slight is the difference between man and brutes. The common person loses this distinguishing feature, while the gentleman (moral person) retains it. Shun (sage king) understood the ways of things and had a keen insight into human relationships. He followed the path of morality.
>
> *(4B19: Lau 1979b)*

These "genuinely human" tendencies that all people share (as well as the psychological consciousness of such potentials) is a source of dignity for the self and the basis of common, or equal, human worth in the Confucian project. Mencius refers to such dignity as "natural nobility" or the "nobility of Heaven" (*tianjue*天爵), which is contrasted with "aristocratic dignity," or what is called the "nobility of man," or "human nobility" (*renjue*人爵). As he explained:

> There is the nobility of Heaven and the nobility of man. Humaneness, rightness, propriety, and wisdom – and taking pleasure in doing good, without ever

wearying of it – this is the nobility of Heaven. The ranks of duke, minister, or high official – this is the nobility of man.

(6A16: Lau 1979b)

Thus, the four moral sprouts of compassion, shame, modesty, and right and wrong endowed by Heaven are the sources of true nobility, which cannot be substituted by human nobility. The nobility of Heaven is absolute and universal to all human beings, as Mencius reminded us: "A sage belongs to the same type as the common people" (2A2: Lau 1979b).

Natural equality forms the basis of a common humanity. It is the kind of equality people have before they mature and assume a certain social role. Hence, what is of real importance is the cultivation of the inner life of the agent and the development of the moral sprouts. The dignity achieved involves more than mere survival and has priority over human nobility.

But Mencius is not simply claiming that merely by having the innate human *qing* or emotions one achieves dignity. What he finds worthy of honor and respect in all human beings is their innate inclination to care in a special way for one another. Each has the capacity to develop such inclinations and gradually to "extend his/her love for those he/she loves to those he/she does not love."

On this understanding, the concept of dignity in the Mencian account has a double structure: first, dignity as inherent and something we recognize as definitional of being "human," and second, dignity as a human good that is earned, cultivated, or achieved.

In its first sense, dignity is grounded in one's intrinsic natural nobility, which is contained within oneself. In its second, dignity refers to honor or esteem achieved or conferred externally through developing oneself to realize one's potentials in the context of social roles and relationships.

IV. Relatedness and equality in Confucian human dignity

Mencius's account of human dignity, as we have seen, emphasizes the process of developing one's moral character in the direction implicit in the predisposition of the heart-mind.

Cheng (誠) is a key concept in the Confucian project of cultivation. The Chinese character for sincerity, constancy, and trustworthiness, it implies consistency among one's feelings, thoughts, words, and deeds.

Thorough cultivation of the internal realm brings about an inner state in which there is *cheng* (誠) in the disposition of the human agent. It refers to an ideal state of being true to oneself, and having no discrepancy, not just between the way one is and one's outward appearance and behavior but also within one's heart-mind, the locus of all cognitive and affective activities.

Leading a moral life requires cultivating our natural moral potentials in the context of human relationships and social roles so as to realize our "natural nobility" and develop our humanity. It is believed that in a truly reciprocal role relationship, where there is mutuality of role performance and interaction based upon reciprocal concern, respect, and esteem, "self" and "other" are constituted, and are constitutive of each other.

The interdependence and interconnectedness between self and others in Confucian moral cultivation explains the paramount importance the Confucians place on the attitude of "*jing*" (敬) or "respect" in the practice of *li* (禮), or rule-following in social life and human interaction. In the classics the word *jing* is used to refer to a frame of mind of serious attention. Confucian *jing* also includes a component of positive regard, one *that leads to a caring attitude.*

From the Confucian perspective, in order to treat people with respect, it is not enough merely to satisfy their basic needs. Human dignity is more than survival. Material well-being and physical security are only the pre-conditions, and are far from being sufficient to support the full development of the natural nobility and moral potentials of humans to fully realize their dignity and humanity, as seen in the *Analects:*

> Tzu-yu子游asked about being filial.
>
> The Master said, "Nowadays for a man to be filial means no more than that he is able to provide his parents with food. Even hounds and horses are, in some way, provided with food. If a man shows no reverence, where is the difference?"
>
> *(2.7: Lau 1979a)*

Here we are told that supporting one's parents is not the most important thing, for one can also support dogs and horses. *Jing* is what distinguishes the two cases. *Jing* or respect refers to an intentional state, involving a mindset and attitude of affective regard and seriousness for the "otherness" of the other. Central to the process of agent self-cultivation through the practice of *li* is the development of one's inner state and sincere emotions to take others seriously and to treat them with genuine regard based on our natural equality.

The intersubjectivity emphasized in the Confucian perspective recognizes that a person's sense of dignity and humanity is also highly dependent on being treated as a worthy individual. The sprout metaphors remind us that seedlings need fertile soil and adequate sun and water to mature. Moral inclinations will wither away if the right nourishment is not given, as Mencius explained: "Given the right nourishment there is nothing that will not grow, and deprived of it there is nothing that will not wither away" (6A8: Lau 1979a).

Highly emphasized in Mencius's theory is the moral responsibility of governments and societies to ensure an educationally and economically nurturing environment, to support the development of our potentials as moral persons. Repeatedly, Mencius reminded kings and rulers of the importance of providing material security to the people so that the heart-mind of the latter will not be overwhelmed by adversity.

At the same time, although all humans are able to realize their inherent dignity as creatures capable of morality, persons can fail to develop that endowment. Even worse, they can turn against their nature. For Confucians, this is what it means to lose one's dignity. Those who give up the effort to develop their moral potentials, thus abandoning what is noble in their nature, degrade themselves and invite disgrace from others.

That is why despite our common endowments and natural equality, Confucius also pointed out that people do develop differently in their course of life through different experiences and circumstances (性相近⊠習相遠). As he said in the *Analects:* "By nature close together, through practice set apart" (17.2: Lau 1979a).

What is expressed here in the Confucian account of human dignity more than two millennia ago is a modern sense of human equality and relatedness – one that importantly allows for both similarity and difference (Bloom 1998: 96).

V. Why harmony requires respect for human dignity

The Confucian project, as we have seen, grounds human dignity in our inborn human *qing* and natural equality. Rather than taking human dignity as simply given, it stresses *achieving* that dignity through self-cultivation. Such a dynamic notion of human dignity and a process view of human agency puts the ultimate control in every person's own hands, as a morally self-responsible agent engaging in a ceaseless process of self-cultivation and transformation.

The notion of natural nobility is an ingredient in the Mencian account of what belongs to recognition of a human being as a *human* being. This natural nobility creates our natural bonding as a human group, and teaches the difference between animals and ourselves. It enables us to recognize our shared humanity, our connectedness, and our natural bonding as the basis of our moral duty towards one another, while at the same time upholding individuality to protect difference and diversity.

Respect for human dignity in the Confucian moral discourse, is a dynamic, two-directional process. The possibility of acting in a morally self-responsible way underpins human dignity and gives rise to the duty of respect. At the same time, one's sense of dignity (the sense of self-respect) and humanity (the capacity to pursue the moral life) develop as a function of one's humanity being treated with respect by others. It is achievable only under socially supportive conditions.

The notion of human dignity as formulated by Kant in his *Groundwork for the Metaphysics of Morals* (1785) is a transcendental concept that is immune to empirical contingency. It is a formal, universal, abstract, and metaphysical attribution of dignity to humanity. As such, dignity does not "belong" to a particular human nature, but is posited in Kant's conception as the basis of the very possibility of human rationality and therefore of morality. Because it is a theoretical abstraction, it cannot be injured or lost to empirical contingency.

The Confucian idea of the dignity of particular persons, however, is vulnerable to empirical contingencies. When people are forced to beg for food due to poverty, to remain silent due to political repression, or to be excluded as non-human or sub-human due to unjust and discriminating government policies, laws, or societal practices, their dignity and humanity will suffer severe damage or injury.

It is therefore unacceptable that, according to the estimation of UNESCO Institute for Statistics, there were over 16.8 million people between 15 and 64 years old in China in 2015 who were still illiterate, among whom some 12 million were female. Notwithstanding the country's wealth, in 2013 over 11% of the Chinese population lived in severe poverty (less than $3 US per day), according to the World Bank Open Data. The social, political, and economic structure of the country is still highly oppressive and exclusive to women, manual workers, peasants, and minority groups in many ways, denying them the opportunities to fully realize their humanity and dignity in accordance with the Confucian ideal.

In a similar vein, we can draw on the moral resources in Confucian universal human dignity to condemn and oppose practices and policies elsewhere in the world that fail to respect our shared humanity or to respond to our common human *qing*, thereby crippling people's inborn capacity to develop and mature as moral persons.

Examples include inhumane governments or societies where policies fail to provide basic education and economic well-being for citizens; where they permit inhumane and cruel practices which endorse honor killing, ethnic cleansing, or torture; or where they promote laws or ideologies that condone the instrumentalization or self-instrumentalization of the human subject, such as sale of human organs, suicide bombing, or certain genetic engineering/enhancement procedures. Equally deplorable are governments and societies that tolerate discriminatory and dehumanizing treatments of human beings, such as racial oppression, stigmatization, and the demonization of other cultures as subhuman or nonhuman.

The intersubjective perspective of Confucian human dignity tells us that we cannot treat others inhumanely without harming our own humanity, nor can we disgrace and dehumanize others without disgracing and dehumanizing ourselves. The humiliation inflicted on us by our perpetrators shows *their* lack of humanity, not *our* lack of dignity (Ni 2014: 182, 184)

This explains why the Holocaust is an account of the loss of dignity and humanity of the Nazi perpetrators who had turned themselves into outcasts of the human community. But their Jewish victims never had their natural nobility taken away, because it is inherent and inalienable. We never fail to recognize them as fellow human beings who share our common humanity.

The same logic applies to the hundreds and thousands of innocent Asian women who were captured by their Japanese aggressors and forced to become comfort women during the war. These women never lost their natural nobility, nor did they turn against their own humanity the way their aggressors did.

Conclusion: Multiple cultures, multiple modernities

Macklin (2003) has suggested that we should get rid of the concept of human dignity, which she claims is vacuous, and instead let the concept of autonomy do all the moral work. What will be lost if dignity is reduced to autonomy?

Mencius's answer would be that we will lose sight of our common human condition, or shared humanity. We will lose sight of our connectedness. Ultimately, we will lose the footing from which to justify why all people should matter.

For Confucians, the special worth of humans is in our ability to go beyond the biologically given by developing and expanding our heart-mind to gradually embrace all things. The Confucian account of human dignity can enable us to recognize difference without losing sight of our shared humanity, can help us accept our mutual responsibilities as morally self-responsible agents, and can provide us with a way to justify why all human beings should matter in the search for common progress. In a world of multiple modernities and intensifying globalization, where pluralism and diversity are a reality, Confucian harmony and dignity enable us to see "connectedness" and to give room for difference. We cannot achieve harmony as a dynamic, generative process without giving consideration to the *due* of each while at the same time recognizing interdependence and valuing difference. We also need to reformulate our ethics to provide a more robust foundation for reciprocal altruism and out-group cooperation that is based on neither strategic calculation nor mere self-interest. A morality of genuine respect for human dignity grounded in natural moral equality and common humanity will prove to be more useful for harmonious living and human flourishing in an age of multiple cultures and multiple modernities.

Note

* ***Author's note:*** Part of this paper was presented as a keynote speech at the international conference on "The Future of Human Dignity" hosted by the Department of Philosophy and Religious Studies of Utrecht University, the Netherlands (October 11–13, 2016).

PART V

Globalization

10

THE NEW MODERNITY

Networked globalization

*Manuel Castells**

Introduction

Modernity is a flexible concept that refers to the rise of a new social configuration that redefines contemporaneity. The modern contrasts with the pre-modern and with the post-modern. Both pre-modern and post-modern are understood in reference to the modern: that is, from our specific perspective in which time/space are defined by the subject, rather than from some chronology external to our experience of a space that transcends our experience. In my view, the concept of modernity is useful precisely because of its ambiguity. Ambiguity is the pre-condition of theoretical innovation, as it leaves the conceptual field, as suggested in Bachelard's epistemology, to the variations of human practice.

Regarded from this vantage point, modernity is always *our* modernity, so the question arises: what is our modernity? What is the modernity characteristic of the early moments of the Common Era's third millennium? I venture the hypothesis that the phenomenon that encapsulates the contemporary experience of humankind at large is what is usually designated as globalization – but not just any kind of globalization. Our globalization, characterized in the terms that I propose here, is networked globalization. In other words, I argue that what defines our form of globalization, our modernity, can be explained as a systemic process based on the "networking of networks" in the practices of actors and institutions.

Globalization: A multilayered network of networks

Globalization is a process characterized by the ability to operate as a unit on a planetary scale, in real time or chosen time, on the basis of the institutional, organizational, and technological capacities installed in the core activities that determine the process and its outcomes.

Institutional capacity means the ability of institutions to induce this process while overcoming political boundaries and regulatory constraints. In practical terms, it refers to liberalization and deregulation. Organizational capacity means the ability of organizations to adopt the form that allows them to operate as a unit in the sense that I mentioned, namely, the networking of operational units. Technological capacity concerns the material ability to work as a unit in real time on a planetary scale, which includes such things as advanced communications and fully computerized telecommunication, transportation, information, and communication systems.

Globalization is multidimensional. It is economic, it is cultural, it is political, it is institutional. Each one of the dimensions is constituted around global networks that connect and disconnect at the same time. The logic of networks is both connection and disconnection. Disconnection is as important as connection, particularly in the case of globalization.

Networks connect people, organizations, activities, and territories from all over the planet according to the goals programmed in each network. At the same time, they disconnect what does not add value to a network, depending again on the interests and goals that are programmed into it.

Since my purpose at this point simply is to illustrate rather than to demonstrate, I will concentrate on the most salient dimension of globalization: the formation of a global economy. I will try to show the critical role of network structure and dynamics in each of the dimensions of the global economy. I will begin with a layer that is most decisive in this case, financial globalization, since it is a key aspect of a capitalist economy, the most pervasive economic form on the planet. In fact, all economies today are capitalist. If capital is global, the core of the planet's economic system is also global.

Global financial markets: Computer networks

As we know, global financial markets are at the heart of the global economy, and they consist of nodes of stock exchange markets, connected by computer networks. Some are physical places like The New York Stock Exchange and The London Stock Exchange, which are highly wired and interconnected within themselves. Other financial markets are purely electronic places like NASDAQ or Eurex. But by now all stock exchanges, whether physical or virtual, are interconnected. So, there is a global network of stock exchanges which is the center of the global economy. These networks process money and securities on the basis of information and flows of instruction. Each network is programmed to ensure that it will attain the goals of financial markets. These goals, in turn, are complicated, but we can boil them down to the essentials.

The primary goal, of course, is to make a profit, and that entails the creation of value achieved through the transfer of value into the financial market. The second goal is to ensure reliable investment conditions. Trust is the foundation of investment, in the sense that it means investors will have enough confidence in the market to put their money into it. The third goal is to maximize the return on investment and thereby to foster the expectation of such returns, which is the driver of investment. The fourth is to accept variations in the process of market valuation as part of the system, hoping then that the market will correct itself and maintain the tendency towards stability so that even if there are ups and downs in equity prices, the market will not be disrupted.

From these general goals, a number of rules and regulations may be derived that are specific to each market. These act as resistors, connectors, and switches in the network, all set differently such that each market has its own particular regulations. Where the flows stop, there are resistors; yet at the same time the network may develop its own connectors to get around those resistors. One such strategy is to invest in hedge funds offshore and then, under favorable conditions, to bring one's investment back into the regulated market. Alternatively, a protocol of communication will be formed between two markets by transforming investments into derivatives that combine the values of commodities and securities in diverse markets. Or, to cite one final example, a switch will be activated by modifying market regulations to build bridges around the blocking action of the resistors noted above.

In fact, much of the process of deregulation and liberalization that took place during the 1990s consisted in reconfiguring financial markets to make them as homogeneous as possible in terms of the conditions under which the flow of capital would operate worldwide. Because such capital flows could now take place in real time in fractions of a second, the system would not be able to operate without mathematical modelling, information systems, and computer networking.

Thus, although the global circulation of money dates back to the beginning of the modern age, this instantaneous flow of transactions around the world based on new forms of communications technology is altogether new.

Moreover, valuations of stocks and securities in general are not the result of strict economic calculation. Analysis has revealed the crucial role of information turbulence in the valuation process, meaning that financial markets are connected to information flows of all kinds (e.g., those in the media), such that information flows generated outside the financial network can fundamentally affect the function of that network. In short, the linkages between media and financial networks, and between political and financial networks, are uncontrolled and uncontrollable. What is reported in the media influences the thinking of investors, and what investors do with this information determines what happens in the financial market.

So, the openness of the financial network, the ability of investors to enter it electronically worldwide, and the accelerating velocity of information turbulences throughout the globe all increase the network's size and complexity, making it uncontrollable in its totality, even if government regulation may control some specific flows under certain circumstances.

In addition, the global financial market is connected to the networks of production and consumption in every country and ultimately determines the livelihoods of ordinary people. In most cases financial markets are not protected by institutional switches (i.e., regulations), yet they affect the ups and downs of the overall economy, and thereby the entire society.

These global financial networks are characterized by flexibility, scalability, and survivability. Their networking capacity vis-à-vis the rest of the economy ensures that they will keep expanding and that their logic will dominate any other sector of the economy, wherever its physical location may be. On the other hand, their openness to information turbulences emerging from other networks introduces static into the information concerning valuation. For that reason, there sometimes will be a mismatch between the networks' internal programming and the external information that they receive from other networks. Those circumstances introduce structural volatility into global financial networks, as we saw during the financial crisis of 2008–12, demonstrating the unsustainability of current information flows.

Global production

The globalization of the production of goods and services is also based on networks. A global production system in fact is a small part of the world economy in terms of employment and numbers of firms. The core of the production system in the world, encompassing both goods and services, is formed by multinational corporations, with their ancillary firms and networks of firms. These are not all Western companies; there is increasing diversity in their national origin, with Chinese and Indian multinationals on the rise. It should be noted that many firms, especially Chinese ones like Huawei, are in fact state-owned enterprises. Ancillary firms are connected by networks within the large corporations and are linked to one another, forming networks of small and medium-sized businesses. Corporations often are decentralized internally; indeed, some of them enter into partnerships that result in networks with other corporations. When this happens, the operating system is formed by internal networks of corporations related to other corporations through expanded networks. The latter, in turn, are tied to small and medium-sized business networks. These networks-within-networks are then linked in various ways to the rest of the economy, although the networked firms remain the leaders.

What is accounted as international trade in its large majority is in fact internal trade to a corporation, networked corporations and their ancillary suppliers shipping their products and components across the borders of different countries, as production is organized in an international assembly line. To be excluded from this network of networks – of production, management, and distribution – means to be excluded from the global economy. But to be excluded from this system does not mean to be unaffected by it, since all economies of all countries ultimately depend on their connection to the components of the economy that are connected to the global networks. For instance, in Latin America, 40 percent of the urban workforce participates in the informal economy, but this informal economy is entirely dependent on what happens in the globalized core of the network. Again, we see a hierarchical structure of networks and their ability to influence what happens in the various territories or activities which are outside of the network.

Globalization of knowledge

A key component of the current global economy is the production and distribution of knowledge and technology. Numerous studies show that knowledge and technology nowadays are produced in global networks of corporations, leading universities, and research centers. There is little scientific research, knowledge production, and technology development outside these networks. These circumstances do not imply that everybody has to be at MIT or Imperial College, but the critical matter for anyone attempting to develop and distribute knowledge is to be part of that global network. And without such a "ticket," you cannot enter it. If you add value to the network, even on a relatively small scale, then you are in. If you do not add value, there will be no admittance into these knowledge-generating networks.

Global markets

Markets of course are connected through the process of reciprocal opening and liberalization under the regulatory environment managed by the WTO. The variable geometry of integration of a given market is a function of how certain key questions are answered: How integrated into the world economy are markets? How open are they, and to what are they open? The status of such markets ultimately determines the wealth and poverty of the nations in which they are located. So then, international trade is also organized around specific networks with different and varying geometries determined through institutional processes.

The global network of networks and metropolitan nodes

The different dimensions of globalization – economic, cultural, and institutional – are organized in specific networks. While each of them has a network of its own, these networks also connect with one another in various ways. The points of connecting points define the spatial architecture of our societies. In other words, all these networks come together into certain nodes, which are network connectors, and this is what defines the spatial structure of our world. This observation explains why, in the age of advanced communication and in spite of all the futurology predictions, we have not witnessed the dispersal of activities across space. On the contrary, we are seeing the largest wave of urbanization in history and the emergence of mega-metropolitan regions. This is because in each country the key connections of all these networks, of all these global networks, come together in one, two, or three major metropolitan areas, depending on the size of the country. Around these connecting points are created opportunities on the basis of wealth, financial production, services, knowledge, communication, and highly skilled labor. In addition, they make it possible for micro-personal networks to reach out and control events from these locations, the macro-functional networks.

In other words, if the few people connecting and deciding on each network are physically together in these few locations, they can manage the network all across the planet through their new communications systems at these spatial connecting points. Around these nuclei of high-value production and control, jobs and opportunities concentrate, and therefore they distribute wealth, expand markets, and induce a self-reinforcing process of metropolitan growth.

Global labor?

To a great extent, labor is also organized by various firms in global networks. On the one hand, there are global networks for "brain circulation" (here we prefer Anna Lee Saxenian's term rather than the more usual "brain drain"). Circulation is a better choice of words here because, in the case of highly skilled labor, one often will find Indian, Chinese, or Brazilian engineers going to Silicon Valley, becoming entrepreneurs and even CEOs, and then returning to their own countries and building new bridges between the two areas. These networks of highly skilled entrepreneurs and engineers, financial analysts, or anyone else who adds economic value are absolutely critical to the functioning of the global economy and indeed of the economy as a whole.

Low skilled, manual labor also plays an essential role, particularly in European societies and even the United States, where there are declining or stable native-born populations due to low birth rates. This low-skilled manual labor is also built around networks and clusters that feed into one another. The latter may include networks of kinship, ethnicity, and curricular affinity, as has been shown by Douglas Massey, Alexander Portes, and a number of other researchers. On the other hand, most labor in the world certainly remains local. Despite the emphasis on migrants in the news, they represent less than 300 million people in the world as a whole, if we count as migrants only those employees who are not citizens of the countries where they live and work. Still, as in the case of the multinational production system, this pool of manual labor, both highly skilled and less skilled, plays a crucial role in the global economy as a whole.

Multi-layered global networks: Theory and practice

This analysis can be extended to other dimensions of globalization, for instance the media. We could analyze in these terms the logic of institutional networking among different states, along the lines proposed by researchers such as David Held, Ulrich Beck, and myself. Such an approach also could be extended to social movements that are challenging the values of globalization, since the globalization process also generates conflict. Indeed, it is not only that certain functions predominate and that certain values pervade the logic along which these global networks are being constructed. Social actors react according to their own interests and values, which means they often will challenge the policies that are scripted in the network. Corporate networks do not operate automatically; instead, they are programmed by deliberate social decisions and choices among values and interests. Hence, there may be resistance to these values and interests, and voices may be raised in favor of alternative projects. To put it differently, the social movements challenging the policies made by the corporate networks also have begun to organize themselves into global networks, as in the worldwide movement against corporate globalization.

This line of reasoning has decisive practical consequences because, in principle, theories and theoretical interpretations carry real-world implications. So we had better be careful in the kinds of theories we construct. For instance, a network analysis of globalization implies that the planet is now organized around networks that both include and exclude. Thus, if particular countries, societies, and people are not included in those networks, or if the programs of these networks are not modified, they will operate according to a nearly automatic logic that will continually reproduce inequality and social exclusion. Thus, either a network will keep on expanding along with the exclusionary logic built into it, or else the programs of the network will have to be altered. This theoretical perspective has practical consequences, e.g., for business, and that is why most major business corporations by now have adopted what they themselves call the network business model. That is, their competitiveness and productivity depend on their ability to network internally and externally with suppliers, customers, financing sources, partners, marketers, and experts in technology. This networking requires organizational restructuring and technological capacity. It also requires a complete retooling of management to avoid what Guy Benveniste terms "coordination errors": that is, the errors that may arise in linking up different networks without taking into consideration the specific codes and rules implicit in each. The critical issue is not so much what used to be called the re-engineering of the corporation. Rather, the really fundamental problem is the re-engineering of the network in relation to the corporation. In fact, the external networking of the company becomes a source of disorder, if internally the company is not networked and thus is unable to assume the autonomy of the node integrating the company and its network.

Knowing that they are in a networked world, workers must put greater emphasis on their flexibility and adaptability. For example, they must learn to develop their professional portfolios so that their skills are, or appear to be, transferable to different business environments. This requires an emphasis on their self-programming capabilities and their ability to manage change.

For labor unions, the implications are staggering. Union density is falling and now mainly concentrated in the public sector. Because unions have waning appeal in the private sector and for the young, government now have become the main employers of unionized workers.

However, unions with foresight are deliberately enhancing their networking. They are building what they refer to as "network unions." The concept of a network union suggests that they will think less in terms of the labor union at the level of the plant, the industrial sector, or even the country. Instead, they will try to reproduce the logic of multinational corporations across national boundaries. They are connecting workers not through the bureaucracies of the labor unions. Rather, starting from plants and organizations, they will try to maintain constant online interaction. This may enable them to become globally networked actors in respect to information and strategy, much as their members already are. Of course, the task confronting labor unions is extraordinarily complicated, since they must transform themselves so completely.

For governments, the trend toward networked globalization means they must focus on developing their internal and external organizations around the notion of shared sovereignty. Certainly, this has become common practice in the European Union and increasingly in other parts of the world and in international relations. However, the networking of nation-states triggers crises concerning the compatibility or incompatibility of their policies. The current crises of the European Union epitomized by Brexit furnish one obvious example. In this context, traditional power politics gradually is being replaced by "noopolitics," a concept introduced by Arquilla and Ronfieldt. As they see it, political strategy today is based on one's ability to influence the ideas and values underlying political decisions, thus influencing the minds of decision-makers rather than relying on traditional power mechanisms to get the decisions one desires.

This does not imply that we will inhabit a world of intelligent decision-making. We still will retain the kind of arrogance that stems from ignorance about our own power. For example, we will keep trying to bomb networks out of existence, but we never truly will succeed in doing so. Or, we may attempt to disrupt communications by introducing disinformation. But political leaders who engage in such old tactics in a new, networked environment do so at their peril, and ultimately fail, although their failures often come at substantial cost to society.

For individuals, the new globalized reality means that they must learn – and are learning – to build their own networks, using information and communications technology beyond their immediate social circles. Their aim, whether or not it is clearly recognized, is to expand the reach and increase the density of their self-created networks, which will enable them to preserve or expand their autonomy vis-à-vis business, institutional, and media networks. These "popular" networks may be individual or collective, but they tend to be as adaptable, flexible, scalable, and dynamic as business networks. For instance, people may seek to set up horizontal communications systems over the Internet and mobile communications in order to win a degree of independence in relation to the media and to controlled sources of information. These trends mark the emergence of what I have conceptualized as mass self-communication, which is of course the most important development in the communications world today.

Finally, a world of networks has not evolved without power and without conflict. Power is exercised through the networks; conflict arises over the programming or reprogramming of the networks. But the whole system appears to be a world without centers. Under such conditions, power – which is ultimately what shapes our societies – is exercised primarily not so much by the conquest of institutions as by actions taken to influence the symbols of society, and the information that drives the economy. The fundamental battle in a networked world is over human minds. This is the historical relevance of what I have conceptualized as networked social movements whose rise we have observed around the world. The May 15

Movement in Spain or the Umbrellas Revolution in Hong Kong come to mind as two relevant examples. Modernity has always been conflict-laden. There is not just one modernity. Social actors with conflicting interests and values seek to define their own versions of modernity, both locally and globally, through their networked practices.

Note

* ***Author's note:*** This essay should be regarded as a rough draft rather than a finished product, especially since the arguments presented here are still tentative and not yet supplemented by appropriate scholarly citations.

11

GLOBALIZATION, MIGRATION, AND THE ROLE OF THE STATE

Julian Nida-Rümelin

Forms of life

We should think of the interactions of human society, both within national borders and beyond them, across the entirety of human history, as resembling a collection of different densities of a myriad of particles in a body of moving liquid. In some places, these particles form dense clumps, while in others they are widely dispersed. Some groups remain close together for longer periods of time, while others spread apart and never encounter one another again. Some particles are highly mobile and only approach one clump or another for a brief moment before moving on to the next. The currents of this liquid are structured by a series of containers, among which particles move at different rates. Patterns of movement and transfer differ, not only from container to container, but also among the individual clumps.

The advantage of this metaphor is that it renders clear and visual two salient features of human society: the gradual and continuous nature of change, and the interaction between structure (the containers) and the particles in motion. The individual clumps represent regions of social intimacy and cultural groups of all different kinds. The process whereby a particle moves from one clump to another and the interactions between different clumps in the flow process represent the processes of creolization and cosmopolitanization within our society. The individual containers are not hermetically separated; they are all interconnected, some more so and some less. There are very few states in the world that can effectively cut themselves off from the continuous process of human migration.

Individual clumps are formed by specific types of interaction that thin out toward the edges and often blend imperceptibly with other interactive structures and their normative representations. It is also possible to discern a shared pattern across all the different clumps and containers. This pattern can be interpreted as an image of a *global civilization*. It is not created by a particular status, but is a continuation of pre-existing connections and interactions that become ever more ephemeral when seen on a global scale that calls for an overarching normative status, over and above particular connections.

In this metaphor, we have the structure of the containers, of interactions, institutionally reinforced collective identities, as seen in the form of a binding and sanctioned rule of law over all cultures and regional associations. We also have the formation of a softer structure, with little or no institutional foundation – that which we would usually call cultural identity. These structural features are not independent, but neither are they identical. An ethnic misconception of nation-states identifies each one with a particular cultural form of life and thereby sees any increase in the diversity of cultural and regional communities as a threat. This ethnic attitude toward a nation-state leads to an inclination to level the cultural playing field, and in situations of conflict to oppress any minority groups and thereby provoke a revolt that uses ethnic identity as a legitimizing factor. This model can be seen in the Kurdish conflict in Turkey as well as in separatist movements such as in Catalonia or in Scotland. The very thing that was supposed to be eradicated by the cultural "levelling" is instead formed into a resistance movement that sees itself as standing for a new "nationality." Instead of allowing the differences in their forms of life to be eradicated slowly under the derogatory term "mountain Turks," as was expected and pursued by the Turkish nationalist movement, the people formed a virtual Kurdish nation that threatened the territorial integrity not only of Turkey, but of three other bordering nations.

Religious communities also follow the same pattern. Communities that feel themselves to be marginalized, that are not permitted to build their own places of worship, that must remain in private and in a certain sense must observe their religious rituals and practices out of sight of the broader public, are presented with a choice: Either assimilate and surrender a part of their identity, or turn this oppression into the core of (possibly imaginary) opposition identity. In this light, the culture in the United States seems to be an excellent example, shaped by a high degree of religious sensitivity and broad range of religious practices. Houses of worship of different faiths, not only Christian, so far have not faced resistance, and the wide variety of Christian faiths reduces the likelihood that members of other faiths will feel that they are being persecuted by a hostile Christian majority.

In non-monotheistic religions, it is far more common to see different religious traditions cohabiting in harmony, sometimes even using the same place of worship or temple complex. Followers of Shinto, Buddhist, Taoist, and Confucian beliefs do not see these systems as irreconcilable. Within these cultural regions, proselytizing followers of different beliefs is not common. We also do not see the close identification between religion and culture that is common in monotheistic religious cultures, one that can even extend to harsh punishments such as the death penalty being imposed on those accused of abandoning the faith (apostasy), as seen in the laws of some Islamic states. The practice of endogamy – that is, of marrying exclusively within one's own religious community – is also historically connected with the three monotheistic desert religions.

In order to have convivial relations between different religious groups, a higher-order consensus is required regarding attitudes and behavior towards those from a different religious or cultural background. This higher-order consensus will call for everyone to treat others with civility, regardless of different value systems. But the practice of civility towards those from different religious and cultural backgrounds also needs to be firmly grounded in everyday culture. A policy of religiously motivated apartheid, with separate meeting places, entertainment locations, and places of work, would be incompatible with this form of civility. Similarly, the practice of dividing large cities into different, internally homogenous religious or cultural districts can initially reduce a certain kind of conflict, while in the long term remaining an obstacle for a shared, citizen-oriented (political) identity.

Liberal philosophers who strive so hard to separate the political from the cultural offer only false hope. We are confronted with a continuum of small differences, starting in the local region of cultural and religious communities and building up to the normative constituents of civilian statehood and global citizenship. The practice of political participation, that is, of republican democracy, indeed can significantly contribute to de-escalating cultural and religious conflicts; however, extending this practice beyond the scope of a nation-state remains a (cosmo-political) postulate.

Global justice

In his most recent monograph, *The Idea of Justice*, Amartya Sen, winner of the Nobel Prize for Economics, warned against describing the ideal status of a just society at the beginning of an experiment. In this, he goes against the most important figure in the concept of justice in the 20th century, John Rawls, who in *A Theory of Justice* (1971) developed the principles of an (almost) just society, but never showed the paths that might lead to an increase in justice in an unjust world. Amartya Sen, both an economist and a philosopher, has been heavily criticized for his attack on Rawls. I believe, however, that most of his critics did not look into the theoretical foundations of his argument against describing a just ideal society. Sen spent decades making significant contributions to the logic of collective decision-making, and thereby pushed the envelope of the optimization model.

The limits of the principle of optimization are seen not only in economics, but also in practical philosophy, in ethics, and most of all in theories of justice. We in fact can describe what would count as an improvement in individual cases, but we cannot conclude from this that it is possible to bring about an optimal situation. In this way, different criteria of improvement can be in conflict, or in sum only permit an incomplete assessment. This leads to the conclusion that it is better to take every opportunity for an (incremental) improvement as it arises, rather than waiting for the opportunity to achieve an ideal status of optimal justice in one fell swoop.

In addition, the history of political theory shows that taking orientation from an ideal goal usually leads to a utopian outlook that is either practically ineffective or involves inhumane conduct. This is certainly true of the first important representative of a utopian viewpoint, Plato, and even more so of his later followers

Tommaso Campanella, Thomas More, and the early socialists of the 19th century. Even Karl Marx, who harshly criticized utopian socialism, is driven by the hope that his own movement would culminate in an earthly paradise. This hope also characterizes his followers in the 19th and 20th centuries. Ideal conceptions of justice often sound like a secularized religious belief, or even a form of political religiosity: They look for salvation in a paradise on Earth instead of in the afterlife, but their schemes remain eschatological in character.

Just as humanism can shade over into utopianism, pragmatism runs the danger of evolving into a technocratic practice. A humanistic position needs to strike a balance between utopia and social technology. Cosmopolitans understand global society as a form of cooperation the fruits of which should be distributed fairly, so that all human beings have the opportunity to lead their lives as they wish (autonomy, humanistic individualism). Cosmopolitanism denies the validity of dividing the world into parcels of nation-states that manage their own internal affairs according to a shared conception of justice, while at the same time fighting like wolves in the global sphere for control of economic resources (workers, raw materials, innovations, etc.). The developments of the last thirty years have reduced the role of individual nations in this battle for resources, and important new players have emerged, such as powerful firms and non-democratic international institutions.

From the cosmo-political perspective, migration cannot be seen by business corporations or destination countries as just one more instrument in the ruthless battle for economic advantage. Well-managed immigration policy that is driven by the economic interests of the destination country can have immense advantages for the economic growth of developed countries, insofar as these countries are experiencing demographic shrinkage. However, the quest for economic benefit must not become the dominant criterion for managing flows of migrants. If immigration policy awards points to migrants based on their usefulness to the destination country, and subsequently offers immigrant status or eventual citizenship, the economic dominance of the northern hemisphere will continue and will even be reinforced by this instrument, cementing the dependence of the southern hemisphere on the northern.

In light of the ongoing dynamics and growth of the global economy and productivity, as well as the scandalously one-sided exploitation of this growth to further the interests of a small percentage of the world's wealthy, I believe that the highest priority must be given to ending the abject suffering of the world's poorest two billion people. Perhaps the steadily increasing pressure of economic mass migration, felt particularly strongly in poorer regions of the world, finally will help to convince the wealthier regions that ending the suffering of the world's poorest inhabitants should be a primary concern for the international community.

The geopolitical paradigm that came into being following the collapse of the Cold War order of two superpowers and was shaped not only by Russia and the USA but also by regional hegemons and the future superpower, China, must yield to the paradigms of global social and domestic politics. I am absolutely convinced that this will require not only the goodwill of individual governments, but also the gradual construction and expansion of *global institutions* that will set the framework within which legal disputes can be rationally argued and practically resolved. The strategy of seeking intergovernmental treaties and agreements has been failing for decades – for example, the sustainability agenda from Rio de Janeiro or the Earth Summits that followed. I see no sense in persisting with this form of global politics, for as far as I can tell, treaties of this kind serve mainly to calm the global masses while practical results remain marginal.

This first, fundamental paradigm shift should be followed by another: namely a shift from a transfer policy of international justice to a regulatory policy. The transfer policy, which can also be seen in social welfare systems of economically developed countries, relies on imposing a charge on economically stronger members of the group and transferring this wealth to the economically weaker. As important as individual transfers of wealth can be, they almost always lead to a situation of dependence and are combined with a loss of agency and, in extreme cases, a loss of authority over the recipient's own life. In developed countries, those who hand over control of their lives in order to gain social welfare benefits lose a considerable amount of autonomy and dignity in the bargain. A rational and just policy within national boundaries should empower and encourage the economically disadvantaged, while fostering their independence and economic viability. Transfer payments should be seen as a measure of last resort and not as the central instrument of control in social policies, and the same principle holds true for international justice as for national welfare programs.

A fair global economy cannot be created by the economically developed countries committing a certain percentage of their GDP to economic aid in the form of money transfers to "developing countries." Such transfers only will lead to a situation where local governments are dependent on handouts from economically developed countries, which among other things leads to the perception that the needs, preferences, and complaints of the region's own population are less important than appealing to the benevolent charity of rich donor countries. This generalization holds especially true for countries without effective democratic rule. Transfer payments also very often have catastrophic consequences for local economic and social structures. For example, they may undermine the local farming economy by providing cheap or free food. In addition, transfer payments tend to perpetuate the kind of structural injustice that the development theoretician Johan Galtung has characterized as structural violence. In many cases, donor countries insist on sealing trade deals for their own national firms and products as a precondition for transfer payments, a situation that only reinforces and entrenches the structure of dependence.

Studies have shown that transfer payments to poorer regions of the world, particularly in Africa, totalling many billions of dollars, more than pay for themselves in the form of new export markets and the perpetuation of structures of dependence. This means that acts that appear to be expressions of charitable generosity are in fact no more than a tool for maintaining the structural injustices that are so beneficial to wealthy countries, businesses, and individuals.

It is interesting to note that the southern hemisphere as a rule profits from periods of political instability in the northern hemisphere. This fact can be seen most clearly in the economic development of South American countries. A fair and just global economy cannot be built out of bilateral trade agreements, but only as part of a gradual multilateral negotiation process run by the United Nations. Such a strategy was successful in the creation of international covenants on rights, and there is no reason to believe it would not be equally successful in creating an international economic order that reflects shared values of justice and fairness. In spite of all conflicts of interest, there remains a common interest in bringing peace and civilization to global society. Many examples from history have shown that the hope of achieving prosperity through cooperation can lead to peace, even in the face of deep-rooted cultural stereotypes and conflicts.

Internationalization

The development of the European Union since the Treaty of Rome in the 1950s has shown in many instances how important it can be to create an institutional cooperative framework that is connected with the expectation of increasing prosperity. The fruitfulness of such a cooperative arrangement is well attested by two of the EU's major accomplishments: the friendship between former "arch-enemies" France and Germany, which took decades to mature, and the sheer inconceivability of military conflict within the EU in spite of countless deep-seated conflicts of interest. Presumably, the Balkan conflicts would never have arisen if this region had been offered prospective membership in the European Community at an earlier point in time. Instead, the foreign policy of Germany and other countries encouraged nationalistically motivated divisions within the former Yugoslavia.

Although we have seen that the two issues of intercontinental migration and poverty in the southern hemisphere are only loosely connected, there is nonetheless a systematic correlation between expectations of prosperity in origin countries and the readiness of their people to migrate. The greater the chances that the local population may achieve economic prosperity by staying home, the less likely it is that they will feel pressure to leave. The new middle classes in poorer regions also have an interest in reducing social tensions, lowering crime rates, and increasing social cohesion in their home countries. But, wherever economic and political events are under the control of a small clique of super-wealthy individuals, that common interest in cooperation and civility will be vitiated. Large regions of South America provide a terrifying example of this problem. When the political system has become nothing more than an instrument to further the interests of a few super-rich families, and when public services are for sale to the highest bidder, politics loses not only all room to maneuver, but economic competitiveness and innovation also are weakened. Relationships solidify and in the end the country is left exploited, looted by an elite who move their wealth and then, often, their families to safety abroad. The economic motivation to preserve public order and encourage social justice plays no role in an oligarchy.

These oligarchies cannot be removed from the outside, but only from the inside, by the political mobilization of voters within the country. But we should be aware that oligarchies all over the world profit from and are stabilized by international trade structures. We should not attempt to export democracy or promote regime change, as the West has done in the Middle East and North Africa region since the beginning of the 1990s, with disastrous consequences. However, we should use our influence to establish global trade structures that place the fulfilment of basic human rights in the foreground, rather than the interests of oligarchs. Direct trade agreements with cooperatives made up of farmers and businessmen should take priority over subsidies. Instituting such agreements will support competitors who eventually can challenge the oligarchy. Companies from economically developed regions who do their business in the Southern Hemisphere can wield that kind of direct influence.

At the moment, local competition among developed countries is blocking effective oversight of their business practices and media strategies. The blockage has given rise to a situation in which some global firms can present themselves as model employers at home while inflating their bottom line by ruthlessly exploiting the inhabitants of poorer regions, requiring them to work under inhuman conditions. That remains as true for the textile industry as it is for model firms like Apple. Here we need an international accord to hold these international companies to ILO standards in a similar way that trade agreements do at a national level. Just as trade agreements contribute in some way to uniformity, non-discrimination, and inclusiveness at a national level, agreements between workers and capital at a transnational, continental, and global level could be agreed upon and policed by a global justice organisation, in a similar fashion to the International Criminal Court. The effectiveness of transnational legal norms within the European Union without the threat of sanctions from an investigative body with powers to investigate what is going on inside member states shows that a similar agreement on humane working conditions is also possible, even without a global police force. The prerequisite for this reform would be a multilateral commitment by all signatory states whereby the established norms would be enforced by local authorities in each country.

The examples given here can be described as routes to a more just society. They include paradigm shifts in collaborative development, the institutionalization of global trade and social policies, the primacy of fighting poverty in the southern hemisphere, political oversight of international companies, and the effort to dismantle oligarchical structures. What role will be played by migration in a world on the path to justice, and what implications would the transition have for the migration policies of the future?

A cosmopolitan ethics of migration

A humane and just world economic and social order will resolve some of the issues that drive migration, but not all of them. Migration as the result of famines caused by misguided global and regional agricultural policy, migration as the result of a hopeless economic outlook in a system controlled by local oligarchs, refugee migration fleeing war and civil war: these forms of migration should decrease under an institutionalized global domestic policy. Members of the United Nations must commit to the rules on avoiding and resolving conflict set out in the UN Charter as well as that body's Universal Declaration of Human Rights and defer to the power of the central treaties and the International Criminal Court. In other words, submitting to the elementary global rule of law constitutes a treaty as the central element of the UN Charter. In order successfully to navigate the path to a fairer international order, the current asymmetry of power must be redressed. As it stands, the only ones who have something to fear from the International Criminal Court are the potentates of powerless states, particularly in Africa, while violations of human rights by powerful countries go unpunished.

On the road to a fairer world, migration to economically developed countries, and in the interests of businesses in these countries, will be permissible only if the disadvantages that are accrued by the countries and economies of origin are compensated in full. War refugees will be given residence in neighboring countries according to a fair distribution formula, financed by the global community, and will be allowed to stay until the conflict has ended. When the conflict cannot be brought to an end in a reasonable amount of time, refugees will be given residence in member countries of the UN, while respecting the preferences of both the refugees and the host countries as far as possible.

Migration between countries at a similar level of economic development will follow the rule: citizens have the (cosmopolitan) right to emigrate, but not the right to immigrate. The country to which these migrants move has a duty of hospitality: it must accept migrants into the country and care for them – in the short term. There is no duty to grant these migrants permanent residency. In realizing this global right to hospitality that Kant described in his 1795 text "Perpetual Peace: A Philosophical Sketch," care must be taken to ensure that those who stay in their home country in spite of a desire to leave are not at a disadvantage. To this end, there should be binding legal mechanisms whereby people who wish to emigrate can apply for asylum in their destination country of choice before departing their own home country. This application could be sent to the (newly-formed for this purpose) local representatives of the UN, or to a UN sub-organization, for example, the UNHCR. Countries can strike bilateral accords on freedom of establishment (a principle of European Union law giving nationals of one country residing in another certain rights, such as that of operating a business) and freedom of movement; alternatively, such agreements can be part of a joint accord among a group of nations. Countries unilaterally can open their borders to migrants, as long as they agree to the rules on compensation for immigrants from poorer regions outlined above. Countries that have a federal structure can set rules to regulate internal migration and have a system of levels of citizenship. For practical reasons, these rules normally will be limited to controls on freedom of establishment and choice of residence location, while leaving freedom within state borders unaffected.

This vision extends the cosmopolitan order both upwards and downwards: sub-national, regional structures will be enhanced and, at the same time, transnational structures will be created. Political organizations will be composed of groups with local, regional, and "national" interests, but also supranational, continental, and global groups, and will practice collective self-determination. A cosmopolitan order of this kind will not replace political and social structures, and it will not turn global society into a market for goods, services, and labor, but rather it will allow the political creation of a just and fair world that permits and encourages desirable mobility while inhibiting the cultural and social damage of mass migration.

12

PLURAL MODERNITY AND NEGOTIATED UNIVERSALS

Jürgen Kocka

Some concepts – "modernity" being one of them – have a tendency to wander, and when they do, they change. Societal projects such as the formation of civil society emerge and seek recognition from specific historical cultures. When they broaden their demands for recognition to include other historical cultures, they also have to transform themselves, or at least they ought to do so. Otherwise, either they will end up not being taken seriously or else they will have to impose themselves by force on the culture from which they seek recognition. In the present age, with its enormous advances in globalization, these reflections should be treated as especially important. Shifts and transformations in the basic concepts of our political and social vocabulary usually go hand in hand with changes in the non-semantic dimensions of historical reality. That, in fact, is what makes it so rewarding to study them. Wandering concepts caught up in the process of expansion and universalization offer instructive lessons concerning the study of mutual social, political, and cultural influences. In other words, they help us examine the encounters between different spaces, or between different parts of the world, e.g., between the West and non-Western areas.

Initially, the concepts "modern," "modernity," and "modernization" bore a European imprint, but over time they became a European–North American hybrid. The word modern and/or its Latin equivalent have been documented since the 5th century. In the 17th century, it became central to the French "quarrel between the ancients and the moderns," and was used frequently in later centuries in various European languages to characterize a range of phenomena from art and poetry to social relationships, types of states, ideas, and entire epochs. In all of those contexts, the term could have three possible meanings. It might be used in the sense of "present-day" or "current" as opposed to "former" or "preceding," or to distinguish the new from the old, or it might mean transitory or ephemeral in contrast to what is eternal. The crucial point here is that all three usages were oriented along a temporal scale. As years passed, what was to be considered "modern" and how it was to be evaluated (in a positive or negative way) was a matter in constant flux. The noun "modernity" came into use around the turn of the 19th century, designating something like the sum total of the most recent social, literary, and artistic currents, again with varying assessments of value.

By contrast, "modernization" did not become a central idea in or outside of the social sciences until the 1950s or 1960s. Strongly influenced by American sources, it contributed to a redefinition of the term "modernity" by treating the latter as a product or touchstone of modernization. The concept of modernization evinced many nuances and shades of meaning, but it also contained some consistent features, of which the following seem especially important. First, like "modern" it was a time-bound concept referring to something new or present-day in contrast to what is old, past, or traditional. To be more precise, the notion of modernization was linked to change; it conceptualized the long transition from traditional to modern ways of life. Second, it marked the next phase of a long tradition in social-scientific and historical theory-formation stretching back to the intellectual world of the Enlightenment in the 18th century. At that early date, theory did not yet employ the notion of modernization; instead, it conceptualized historical change using words such as progress, revolution, civilization, rationalization, social differentiation, or simply history (in the singular) as a unidirectional, irreversible process. Third, "modernization" aspired to be a systematic concept in the sense that it assumed or presupposed an integral process of transformation embracing economic, political, social, and cultural aspects. It was intended to highlight the correlations among phenomena such as the spread of capitalism, industrialization, the crystallization of liberal-democratic structures, the rise of the nation-state and a pluralist society, the ways in which achievement criteria were employed to structure social relationships, the progress of science, and the emergence of certain personality structures, belief systems, and mental dispositions. The notion of modernization also was intended to show that all these correlations were "normal." Of course, various authors, employing different languages, emphasized distinct aspects, but almost all underscored the mutual conditionality of whatever phenomena they had chosen to emphasize. Fourth, the assumption was frequently made (and most clearly articulated by Marx) that the more advanced countries might anticipate the development of the more backward ones, regardless of the undeniable differences among them in matters of detail. The world was destined to become more homogeneous as time went on. And fifth, the tendency just described generally was deemed to be something positive and desirable that ought to be welcomed. Max Weber rightly is seen as the most notable pioneer of this approach, although he rarely spoke of "modernization" and never lost sight of its profound ambivalence.

All of these concepts are unambiguously Western. As social historians we could try to reconstruct the Western European and North American experiences and their underlying structures that suggested this way of thinking and made it seem plausible, at least to some segments of the elite. But independently of these Western origins, the modernization mindset had acquired global dimensions since the beginning of the 18th century. For one thing, in principle the idea of modernization could treat all of humanity as the recipient of its benefits, to be united and led peacefully toward a better future, beginning with its center in Europe. Although the viewpoint may have been Eurocentric, the aspiration was universal. When that vision was put into practice, especially in the French Revolution, it had repercussions on a world-wide scale. For another, when the European and later North American elites invented and articulated this Eurocentric or Western way of thinking, it was part of an emergent European and/or European/North American identity. The latter was formed, among other factors, by encounters with non-European (non-Western) cultures, from which the Europeans (Westerners) distanced themselves: the so-called Orient in the 18th and 19th centuries and the East and "developing countries" in the 20th century. The two world wars also played a role as the settings for fateful encounters. In other words, in retrospect it turns out that the most strongly Eurocentric ideas were, in part, the outcomes of transnational interrelations that extended beyond Europe and/or the West.

Theories of modernization have taken a great deal of critical flak since the 1960s, when they really consolidated their domination of the field. There is no reason to rehearse this critique exhaustively here, but two of the objections against it do indeed carry some weight.

First, classical modernization theories do not sufficiently take into account the catastrophes of the 20th century, a fact that presumably has to do with American power as well as the victorious mood that prevailed after 1945 and was disseminated thereafter in the intellectual sphere. It took quite a while until European authors emerged who were able to accentuate, within modernization theory, the century's traumatic experiences. When that point was finally reached, the optimistic world view of classical modernization theory suffered a severe shock. In this context we should think of Foucault and the reinterpretation of Max Weber, Zygmunt Bauman, and Norbert Elias, the last a writer who is open to both positive and negative readings of modernization. The work of these authors directed attention to the crises and costs of modernization and diluted their teleological implications. The reorientation that followed their critiques included the re-evaluation of the great dictatorships of the 20th century as modern phenomena as well as the discovery of the environmental crisis. Notions of modernization began to change. With ever greater resolve, critics highlighted the ambivalences and dark sides of modernity, the price of progress and the losses incurred by it.

Second, the dissemination and application of modernization theories to areas of the world where they did not originate also helped make them targets of critique. Encounters between Western and non-Western parts of the world, both in academic settings and in other arenas, became far more frequent as they were given impetus by the most recent advances of globalization. These encounters led many to question the assumption, so prevalent in modernization theory, that the world was becoming more homogeneous. Moreover, in-depth comparative historical studies fostered a dawning awareness that growing interdependence did not necessarily mean increasing cultural assimilation. It also turned out that approaches to the politics of development that were indebted to the idea of modernization gradually lost their allure. Ultimately, scholars could not get past the recognition that many traditions do not simply disappear under the impact of modernization; instead, they live on in a modified form and play some role in setting its course. Tradition helps to determine which modern elements will be selected, reinterpreted, and appropriated by the recipient culture. Observations of the process of modernization in non-Western parts of the world – not to mention in non-Western parts of Europe after 1990 – have revealed the decisive role that culture played (and still plays) in influencing the prospects, the failure, and the course of political and economic modernization. In the final analysis, the notion of "embedding" as well as a methodology "enlightened" by the influences of ethnology and/or anthropology shored up the objections to classical modernization theories, which were in any case sometimes too schematic.

But on the other hand, there also have been events and trends that might be construed as evidence in favor of these theories. The collapse of communism and the breakup of the Soviet Union could be understood as confirmations of the thesis that, over the long haul, modern technological change and industrial growth fit together best with a constitutional, non-dictatorial government, the rule of law, and a more or less open society. Without those complementary institutions, technological advancement and growth may slow to a crawl or simply fade away. Modernization theories emphasized this reciprocal relationship. There are other examples that can be explained cogently by the classical modernization paradigm. The case of China appears relevant in this context. The transformation taking place there, wherever it may lead, could reconfirm the view that in the long run a rapidly modernizing economy will be dependent on social and economic modernization, whatever form they may take. Thus, regardless of the harsh, decades-long critique of the concept of modernization, there are sound reasons not to discard it.

To be sure, that notion has changed and continues to change. It has lost its triumphal tone and grown more modest, both substantively and methodologically. Today, people prefer to speak of modernity rather than modernization, which can signify a loss of precision and a relativizing of the historical dimension. However, this shift reflects a stronger emphasis on culture and a more open horizon of expectations. Even more important has been the move from singular to plural: i.e., toward the idea of "plural modernity" or of "multiple modernities," which has quickly gained acceptance. The origins of this conceptual shift can be traced back to Shmuel Eisenstadt. The idea here is that increasing globalization may mean growing interdependence, but need not also entail greater homogeneity. Thus, the expectation that societies will converge is abandoned or at any rate dialed back, while the notions of modernization and Westernization are clearly distinguished. No one country and no single region of the planet should establish criteria for the modernity of the rest.

Nevertheless, this category shift – the pluralization of modernity in the course of its global diffusion – has a cost. The common denominator of all these modernities, the indispensable defining feature that justifies the employment of any concept whatever, whether in the plural or not, is frequently left vague or vanishingly small. For example, the Turkish scholar Nilüfer Göle identifies the chief characteristic of modernity simply as its capacity for constant self-correction. For the Swedish social scientist Björn Wittrock, modernity is neither a unitary civilization nor an epoch; instead it is a global condition, an ensemble of hopes and expectations required only to meet certain minimum conditions of appropriateness. Among other such criteria, Wittrock mentions a specific consciousness of history, a new understanding of the thinking, acting self and its place, as well as the self-reflective, critical nature of modern thought. For his part, Eisenstadt sees a civilization as modern if its order is no longer taken for granted, but instead has become the object of constant contention, as when protest movements have come to play a crucial role and new forms of political legitimation have emerged. Here, too, we notice the absence of economic or political-institutional criteria for modernity. But on the basis of such soft definitions, it is scarcely possible to distinguish different degrees of modernity. At the same time, the dichotomous terminology of modern versus traditional is explicitly relativized or even abandoned. Consequently, almost anything can pass for modern. Yet, as we know, there is not much analytical utility in concepts that include much and exclude little. Of course, that does not imply that there are no diplomatic uses for such concepts.

Still, there are some gains to be realized in the paradigm shift from singular to plural. The idea of the plural nature of modernity/modernities enables us to recognize and acknowledge existing plurality without having to relinquish entirely a unitary reference point. It establishes a framework within which institutions, strategies, and values can be compared with one another in respect to their similarities and differences. Furthermore, it invites us to examine reciprocal influences, the impact of encounters on all the participants, processes of mutual perception, choice, and appropriation, and the complications inherent in the latter. The integration of the modern world finds expression here, and the hybrid character of numerous phenomena comes to light. The plurality of modernity keeps the representatives of Western cultures from overvaluing their own traditions and opens up new, productive possibilities for intellectuals in other corners of the world to put their own experiences and traditions in a global context. They can do so without having to seal themselves off from modernity as such, effectively isolating themselves in an intellectually fragmented global landscape.

This is the point I want to make in speaking of "negotiated universals." It is widely understood that the philosophy of modernity, rooted as it is in the Enlightenment, claims universal applicability and global validity despite its particularistic and regional (European and/or Western) origins. Over the centuries several paths have been available to it for making good on this universal claim, at least to some degree. First, we should include strategies of coercion and domination: universalization via pressure, manipulation, and violence. There have been many examples of this sort of thing in the history of colonialism and imperialism, even up to this very day. Second, the attractiveness of the modern project and the desire to imitate aspects of it have played a role, both past and present. The cultures, political systems, rights and values, and even the economic structures of the West have proved attractive in many respects to other parts of the world, especially to the elites there, which frequently have adopted them lock, stock, and barrel. It must be admitted that these two strategies of universalization – imposition by force and imitation of models – cannot always be kept strictly separate.

A third path, one that offers all participating sides the opportunity to make their own contributions – indeed practically invites them to do so – is that of negotiation. The word "negotiate" is used here in a metaphorical sense. The process it is intended to describe can assume a variety of forms, display many facets, and take place across numerous spheres of life. The crucial point is that the expansion of the spatial and cultural scope of a conceptual scheme or theory, a social project or program, should be linked to a substantive change in the concept, social project, or program in question. The European/Western paradigm of modernization itself can serve as an example here. On the one hand, it is not merely a case of Western particularism; it claims more extensive recognition, aspires toward universalization, and has much to offer other parts of the world today, just as in the past. At any rate it continues to advance and spread. On the other hand – and like other significant schemes, theories, social projects, and programs of all kinds – to some extent it really is context-dependent, culturally specific, or in other words "historical." Hence, modernization cannot simply be exported and implanted in other cultures without either doing violence to them, or allowing the entire project to fail. As conceptual schemes are expanded, transplanted, and appropriated, they simultaneously must be selected, reinterpreted, modified, adapted and inserted into new contexts. The question of whether such a mobile concept is capable of being changed, adapted, and integrated in this manner constitutes one of the quality control checks on this export item. Fundamentalist convictions will fail the test, while enlightened schemes and programs can pass it in principle, because they include a built-in learning curve. But, of course, this is not a purely intellectual issue; it is likewise a question of practice, power, and skill.

Negotiation is a matter of mutual give-and-take. It takes place across distinct spheres of life, for example in the case of science, which clearly is now being practiced at the global level: debates over the goals of knowledge; over which questions need to be raised; over theories, concepts, research designs, and forms of presentation. Such negotiation is also evident both implicitly and indirectly or gradually in the ways in which science is actually done in practice, to the extent that scientists or scholars from different regions of the globe have an interest in cooperation and mutual learning that transcends national boundaries. At stake here are long-term processes of consensus-building for which the institutional conditions must be created deliberately, since they often do not yet exist.

There are at least two aspects of the negotiation process that need to be considered. On the side of the recipients, both in science and in public discussion and politics, a process of partial rejection, selective acceptance, reinterpretation, and incorporation-with-modifications takes place that usually involves conflicts, winners and losers, acts of destruction and new beginnings. One might think of the reception of

Western models in Japan since the Meiji era, which was always only partial. There also will be – or ought to be – some feedback for the originators. Thus, the Western paradigm of modernization is challenged by new discoveries and competitors as it spreads to East Asia, India, or the Islamic world. Self-criticism and self-relativizing are called for in the originating societies, and the conceptual scheme evolves. In the process conflicts may arise among the originators themselves. There are winners and losers; many positions and convictions must be abandoned as new ones crystallize out.

The paradigm shift from classical modernization theories toward the plurality-of-modernities position offers a particularly good example of what is intended here by the term "negotiate." Because negotiation causes a basic conceptual scheme to be modified, its openness to diversity increases along with its capacity for renewal. At the same time, it becomes possible to avoid postmodern fragmentation, mutual disregard, and compartmentalization, while the universalistic elements of the theory are retained or incorporated into practice. This is a process, then, that is not exclusively intellectual; it also has a practical dimension. When it succeeds – and it doesn't always – at its (necessarily provisional) end-point we will find "negotiated universals."

Enough has been said about the global diffusion of European conceptual schemes, Western outlooks, and Occidental practices together with the transformations they undergo as a result. What about a process that goes in the opposite direction? Which concepts, schemes, paradigms, and social projects from East Asia or the Islamic world lay claim today (or once did) to a similarly universal validity, and thus would be suitable for a comparable process of negotiation and diffusion-cum-change? Until the 18th century, Europe tended to be on the recipient side, but in the last two centuries the negotiation of universals has been a highly asymmetrical affair that emanated more strongly from the West than vice versa. After the globalization of the last few decades got into full swing, the balance began to shift toward Asia, or at least East Asia, as the economic and power-political advantages of the West diminished. Thus, one ought to expect more equilibrium both in terms of political ideas and conceptual schemes. This trend increases opportunities for processes of mutual negotiation to occur, while also holding out the prospect of interesting, albeit unforeseeable outcomes. When it comes to the history of ideas and theories, globalization has barely begun.

CONCLUSION

Thomas Meyer and José Luís de Sales Marques

The papers collected in this volume share two objectives. For one thing they seek to clarify the concept of multiple modernities, especially the question of what constitutes its common core and what procedures would best enable us to identify that core. This project assumes particular urgency if one takes seriously the premise that the age of Western dominance in the field of modernization is over. For another, they address the issue of how scholars can apply the concept of multiple modernities to the political realities of today's world.

The papers presented at the conference, along with subsequent discussions with the authors hosted by the Institute of European Studies of Macau, affirm that the multiple modernities approach is analytically fruitful and normatively relevant. However, a pair of objections call for further scholarly clarification. The first concerns whether there are roads to modernization beyond the Western and, strictly speaking, "liberal" modernity that make self-sustaining economic progress possible, whatever else their merits and benefits for the countries in question might be. The second concerns whether the common core of alternative roads to modernization – especially in their political dimension – can be conceptualized with sufficient precision to save it from the "slippery slope" toward relativism.

Shmuel Eisenstadt, who pioneered the idea of multiple modernities, held that it does not stipulate anything close to Samuel Huntington's relativism. Instead, he argued, it presupposes a common core shared by all the different types of modernity. In his view, modernism embraces an idea of the future characterized by various alternatives that are realizable through autonomous human agency, or the principle of subjectivism. In modern societies, the premises on which the social, ontological, and political order is based, and their legitimation, are no longer taken for granted as objective or given. For Eisenstadt, modernization does not necessarily mean Westernization; the Western model of modernization is merely prior in time to other models, though for some of the latter it may still enjoy normative relevance.

This criterion may acquire sufficient explanatory power and conceptual rigor to define both the commonalities and differences among variants of modernization if its political dimension can be made more concrete. Some of the contributors to this book have suggested ways to do so.

It may be the case today that, given the diversity of its regional cultural context, the modernization process can no longer be identified with Westernization in the strict sense of the word, Nevertheless, one might also argue, as Jack Snyder did in his essay, that enduring economic progress seems impossible without the central element of liberalism, or at least a workable substitute for it. From that vantage point, the many roads to modernization may indeed run parallel for a long time, but in the end, they all will need to converge on a liberal platform.

From a Confucian perspective, arguments can be marshalled to refute this objection. A thoroughgoing examination of the central values of Confucianism reveals that they possess a complexity and depth that tends to be underestimated in the West, and in essential respects offer functional equivalents for crucial aspects of liberalism. Above all, that holds true for the socio-political dimension of the principle of harmony developed by Confucius and Mencius. Some have suggested that this principle, as interpreted by the Confucian tradition, offers a way for quasi-theocratic authorities to grant themselves the power to homogenize or suppress individuality and difference, two of the hallmarks of modernity. But as Julia Tao's essay pointed out, the contrary is true: Harmony in their meaning of the word presupposes difference and individuality, and aims to preserve and maintain them by seeking a balance, insuring that justice will be done both to elements of diversity and to the interests of the community as a whole. Any regime whose public posture and policies followed the maxim that individuality and difference should be leveled down would lose legitimacy in regions of the world shaped by Confucian culture. Of course, liberals could raise the obvious objection that individuality and difference will be recognized fully only if they possess a secure institutional foundation. But advocates of this tradition would respond that the highly formalized system of state-administered examinations for public officials in the Chinese tradition encompasses precisely the effective institutionalization that liberals demand. Moreover, two hitherto unanswered questions should be asked of the Confucian tradition's critics. First, does economic progress automatically give rise to a society with humane values and organization? Second, what methods might liberalism adopt to generate the measure of solidarity without which no society or economy can thrive?

If we inquire whether non-liberal paths to modernization really are sustainable, the analysis and assessment of the special characteristics of Chinese capitalism plays an exemplary role. The latter differs substantially from both types of capitalism found in Western countries and highlighted by Hall and Soskice in their "varieties of capitalism" approach (which also takes the cultural dimension into account): the *liberal* type and the *coordinated market economy.* In the Chinese variant, both socio-economic networks and the state – through its multiple functions as market referee, investor, and player – are allotted prominent roles. However, neither the state nor the networks nullify the competitive struggle among the various players.

What happens in the market is influenced, albeit not canceled, by diverse "logics" peculiar to the Chinese cultural environment. Obviously, this economic model is highly successful and, in principle, it shows no signs of having reached its limits. Its combination of acknowledged success and embedding in the main cultural traditions of China may guarantee its long-term sustainability, as Tak-Wing Ngo suggested in his essay. Thus, the case of China constitutes an interesting piece of evidence in favor of the analytic fruitfulness of the multiple modernities model, though that subject requires much more scrutiny.

At the Macau conference, the idea of employing the notions of *good governance* and *human security* to flesh out the political dimension of the common core fell on fertile soil. However, it is unclear whether participants gave that idea unqualified assent. Both concepts reflect the intention to develop globally valid criteria for evaluating political actions and constitutions: ones that cannot be attributed exclusively to the Western tradition, but rather prove applicable to all the variants of modernization. The concept of good governance, like that of multiple modernities, deliberately leaves open the issue of specific forms in which it can be institutionalized. It sheds light on both of the dimensions of government action: input (participation) and output (the provision of individual and public goods). The common core that binds together the divergent paths to modernization is the obligation of governments to seek popular legitimation for their actions through the principle of subjectivity. In the final analysis, that obligation means that the consent of the governed must be solicited, and the request for that consent must respect the equal voice and thus the equal value of every citizen. The concept of good governance ties into that principle of consent, and implicitly raises a normative claim that recalls the Covenants on Human Rights enacted in 1966 by the United Nations. Those covenants affirmed the inseparability of two principles: the rights of political participation and access to social and economic goods. Only when both are universally available can we say that the requirements of dignity and freedom have been met. Adopting this criterion allows us to broaden debates about good government and to make valid comparisons between different forms of government without privileging distinctly Western models of modernization. Of course, identifying subjectivity as the core of modernization via the ideas of good governance and human security could run the risk of misuse. Because there is still a certain degree of vagueness and generality that continues to prevail in the debates concerning both concepts, it might prove true that good governance and human security could be interpreted in rather arbitrary ways to support whatever conclusion one desired.

Thus, despite broad scholarly acceptance of the multiple modernities model, attempts to define the common core within divergent trajectories of modernization remain controversial, especially for social scientists not committed to the Western tradition. In particular, three issues must be elucidated. First, countries currently move along at different rates of modernization in respect to their political constitutions and political economy. Will these divergent paths converge over time (perhaps a very long time)? Or will those divergences become permanent? Second, given that countries in fact have interpreted and institutionalized the common core of modernity in a variety of ways, will those differences meet with reciprocal recognition, and will the content of modernity's core lend itself sufficiently to empirical study? Finally, the question raised by Jürgen Kocka must be clarified. If the common core underlying different paths to modernity should be defined through a dialogue among its proponents rather than taken for granted as an artifact of Western dominance, what implications will follow for practical politics and for political research?

Those are just a few of the challenges that affect not only the way in which the contemporary world is understood, but also the degree to which its political actors will be disposed to cooperate. The essays included in this volume are intended to contribute to a discussion about the political dimension of multiple modernities.

REFERENCES

Peer-reviewed publications

Acemoglu, D. & J. Robinson, 2006. *Economic Origins of Dictatorship and Democracy*. New York: Cambridge University Press

Acemoglu, D. & J. Robinson, 2012. *Why Nations Fail: The origins of power, prosperity, and poverty*. New York: Crown

Acharya, A. 2001a "Human Security: East versus West." In *International Journal*, 56(3), pp. 442–460. Translated into Chinese and Korean

Acharya, A. 2001b. *Constructing a Security Community in Southeast Asia: ASEAN and the problem of regional order*. New York: Routledge

Acharya, A. 2009. *Whose Ideas Matter? Agency and power in Asian regionalism*. Ithaca, N.Y.: Cornell University Press

Acharya, A. 2011. Available at www.un.org/en/ga/president/65/initiatives/Human%20Security/Amitav%20Acharya%20UNGA%20Human%20Security%20Debate%20Presentation.pdf

Acharya, A. 2013a "The R2P and Norm Diffusion: Towards a framework of norm circulation." In *Global Responsibility to Protect* 5(4), pp. 466–479

Acharya, A. 2013b. *The Making of Southeast Asia*. Ithaca, N.Y.: Cornell University Press

Acharya, A. 2014. *The End of the American World Order*. New York: Polity Press

Acharya, A. 2016a. "American Primacy in a Multiplex World." In *The National Interest*, September 27. Available at http://nationalinterest.org/feature/american-primacy-multiplex-world-17841?page=show

Acharya, A. 2016b. "'Idea-shift': How ideas from the Rest are reshaping global order." In *Third World Quarterly* 37(7), pp. 1156–1170

Acharya, A. 2017a. "Donald Trump as President: Does it mark a rise of illiberal globalism?" In *YaleGlobal*, January 22. Available at http://yaleglobal.yale.edu/content/donald-trump-president-does-it-mark-rise-illiberal-globalism

Acharya, A. 2017b. "The emerging powers can be saviours of the global liberal order." *Financial Times*, January 19. Available at www.ft.com/content/f175b378-dd79-11e6-86ac-f253db7791c6 In Chinese: www.ftchinese.com/story/001071070/ce

Acharya, A. 2017c. "Human Security." In J. Baylis & S. Smith (eds.), *The Globalization of World Politics: An introduction to international relations*. 7th edition. Oxford: Oxford University Press, pp. 480–496

Acharya, A. 2017d. "Global Governance in a Multiplex World." *EUI (European University Institute) Working Papers, 2017/29*

Agrafiotis, I., M. Bada, P. Cornish, S. Creese, M. Goldsmith, E. Ignatuschtschenko, T. Roberts, & D. Upton. 2016. *Cyber Harm: Concepts, taxonomy and measurement*. Saïd Business School RP2016-23. Oxford: Saïd Business School, Oxford University

Akca, I., A. Bekman, & B. A. Ozden. 2013. *Turkey Reframed: Constituting neoliberal hegemony*. London: Pluto

Albert, M. 1993. *Capitalism against Capitalism*. London: Whurr

Almond, G. & S. Verba. 1963. *The Civic Culture: Political attitudes and democracy in five nations*. Boston: Little, Brown

Arendt, H. 1973. *The Origins of Totalitarianism*. New York: Harcourt Brace & Jovanovich

Arnason, J. 2002. "Communism and Modernity." In S. Eisenstadt (ed.), *Multiple Modernities*, pp. 61–90

Aron, H. 2016. "Monopolizing the Nation: Evidence from the Israeli Settlement issue on right wing capture of national lexicon and symbols." Working paper

Aron, R. 1967. *Les Etapes de la Penseé Sociologique: Montesquieu, Comte, Marx, Tocqueville, Durkheim, Pareto, Weber*. Paris: Gallimard.

Barbero, A. 2007. *I barbari. Immigranti, profughi, deportati nell'impero romano*, Bari/Roma: Laterza

Bartels, L. M. 2005. "Homer Gets a Tax Cut: Inequality and public policy in the American mind." In *Perspectives on Politics* 3(1), pp. 15–32

Beck, U. & E. Grande. 2007. *Cosmopolitan Europe*. Cambridge: Polity Press

Beck, U. & E. Grande. 2010. "Varieties of Second Modernity: The cosmopolitan turn in social and political theory and research." In *British Journal of Sociology* 61(3), pp. 409–443

Beck, U. 2006. *The Cosmopolitan Vision*. Cambridge: Polity Press

Bell, D. 2015. *The China Model: Political meritocracy and the limits of democracy*. Princeton: Princeton University Press

Benhabib, S. 2002. *The Claims of Culture: Equality and diversity in the global era*. Princeton, N.J.: Princeton University Press

Berman, S. 2006. *The Primacy of Politics: Social democracy and the making of Europe's twentieth century*. Cambridge: Cambridge University Press

Blank, S. & Y. Kim, Y. 2016. "Economic Warfare à la Russe: The energy weapon and Russian national security strategy." In *The Journal of East Asian Affairs* 30(1), pp. 1–39

Bloom, I. 1998. "Mencius and Human Rights." In W. T. De Bary and T. Weimung (eds.), *Confucianism and Human Rights*. New York: Columbia University Press, pp. 94–116

Blyth, M. 2002. *Great Transformations: Economic ideas and institutional change in the twentieth century*. Cambridge: Cambridge University Press

Bodin, J. 1566. *Methodus ad Facilem Historiarum Cognitione*. Paris

Boix, C. 2015. *Political Order and Inequality*. New York: Cambridge University Press

Bowman, J. 2015. *Cosmoipolitan Justice: The Axial Age, multiple modernities, and the secular turn*. New York: Springer

Boyer, R. 1997. "The Variety and Unequal Performance of Really Existing Markets: Farewell to Doctor Pangloss?" In J. R. Hollingsworth & R. Boyer (eds.), *Contemporary Capitalism: The embeddedness of institutions*. Cambridge: Cambridge University Press, pp. 55–93

Boyer, R. 2005. "How and Why Capitalisms Differ." In *Economy and Society* 34(4), pp. 509–557

Boyer, R. 2011. "Are There Laws of Motion of Capitalism?" In *Socio-Economic Review* 9, pp. 59–81
Brague, R. 1992. *Europe, la Voie Romaine*. Paris: Criterion
Braudel, F. 1966. *The Mediterranean and the Mediterranean World in the Age of Philip II*. 2 vols., 2nd revised ed. Berkeley: University of California Press
Braudel, F. 1982. *The Wheels of Commerce*. New York: Harper & Row
Braudel, F. 2002. *The Mediterranean and the Ancient World*. London: Penguin
Bremmer, I. 2012. *Every Nation for Itself: Winners and losers in a G-Zero world*. New York: Portfolio/Penguin
Buckley, P., J. Clegg, & H. Tan. 2005. "Reform and Restructuring in a Chinese State-Owned Enterprise: Sinotrans in the 1990s." In *Management International Review* 45(2), pp. 147–172
Buffon, G-L. 1761. *Histoire naturelle*. Paris
Caporaso, J. & S. Tarrow. 2009. "Polanyi in Brussels: Supranational institutions and the transnational embedding of markets." In *International Organization* 63(4), pp. 593–620
Cardini, F. 1994. *Noi e l'Islam: Un incontro possibile?* Bari: Laterza
Carr, E.H. 1958–1964. *Socialism in One Country, 1924–1926*. London: Macmillan
Cerutti, F. et al. (eds.), 2010. *Debating Political Identity in the EU*. London: Routledge
Chen, Z. 2013. "The Efficiency of Post-Lisbon Treaty EU's External Action and China-EU Strategic Partnership." In M. Telò & F. Ponajert (eds.), *The EU's Foreign Policy*. Burlington: Ashgate, pp. 175–187
Cheng, H. & Y. Cheng. 1981. *Er Cheng ji*. Beijing: Zhoughua shuju
Cheng, Y.-W. & T.-W. Ngo. 2014. "The Heterodoxy of Governance under Decentralization: Rent-seeking politics in China's tobacco growing areas." In *Journal of Contemporary Asia* 44(2), pp. 221–240
Chinn, M. D. & J. A. Frieden. 2011. *Lost Decades: The making of America's debt crisis and long recovery*. New York: Norton
Cooley, A. 2017. *Dictators without Borders*. New Haven: Yale University Press
Cooper, R. 2000. *The Post-Modern State and the World Order*. New York: Demos
Cox, R. 1996. "Towards a Post-Hegemonic Conceptualization of World Order: Reflections on the relevance of Ibn Khaldun." In R. Cox with T. Sinclair (eds.), *Approaches to World Order*. Cambridge: Cambridge University Press, pp. 144–173
Dancygier, R. M. 2010. *Immigration and Conflict in Europe*. New York: Cambridge University Press
de Bellaigue, C. 2016. "Welcome to Demokrasi: How Erdogan got more popular than ever." In *The Guardian*. August 30, 2016
de Las Casas, B. 1992. *Historia Apologética*. Madrid: Front Alianza
de Pauw, C. 1768. *Recherches philosophiques sur les Américains*. London
de Pauw, C. 1771. *Recherche Philosophiques sur les Américains*. London
Dennison, S. & D. Pardis. 2016. *The World According to Europe's Insurgent Parties: Putin, migration and people power*. European Council on Foreign Relations. ecfr.eu, June 27
Diamond, L., M. F. Plattner, & C. Walker (eds.) 2016. *Authoritarianism Goes Global*. Baltimore: Johns Hopkins University Press
Dollar, D. 2015. "Institutional Quality and Growth Traps." *Pacific Trade and Development Working Paper Series*, No. YF37-07. Presented at the conference of the Institute of Southeast Asian Studies, Singapore
Donnelly, J. 2003. *Universal Human Rights in Theory and Practice*. 2nd edition. Ithaca, N.Y.: Cornell University Press
Drezner, D. W. 2014. *The System Worked: How the world stopped another Great Depression*. New York: Oxford University Press.
Durkheim, E. 1965. *The Division of Labor in Society*. New York: Free Press

Eichengreen, B., D. Park, & K. Shin. 2013. "Growth Slowdowns Redux: New evidence on the middle-income trap." *NBER Working Paper*, No. 18673

Eickelman, D. 2002. "Islam and the Languages of Modernity." In S. Eisenstadt (ed.), *Multiple Modernities*, pp. 119–135

Eisenstadt, S. 2000. "Multiple Modernities." In *Daedalus* 129(1), pp. 1–29

Eisenstadt, S. (ed.). 2002. *Multiple Modernities*. New Brunswick, N.J. & London: Transaction Publishers

Estlund, C. 2017. *A New Deal for China's Workers*. Cambridge: Harvard University Press

Evans, J. A. 2005. "The Dynamics of Social Change in Radical Right-Wing Populist Party Organization." In *Comparative European Politics* 3(1), pp. 76–101

Fairbairn, B. 1997. *Democracy in the Undemocratic State: The German Reichstag elections of 1898 and 1903*. Toronto: University of Toronto Press

Ferrant, G. & A. Kolev. 2016. "The economic cost of gender-based discrimination in social institutions." *OECD Development Centre Issues Paper*. Available at www.oecd.org/dev/development-gender/SIGI_cost_final.pdf/

Fitriani, E. 2015. "ASEAN and the EU Cooperative Culture in ASEM." In M. Telò, L. Fawcett, & F. Ponjaert (eds.), *Interregionalism and the EU*. Farnham: Ashgate, pp. 249–265

Frank, T. 2004. *What's the Matter with Kansas*. New York: Henry Holt

Frieden, J. 2016. "The International Governance of Finance." In *Annual Review of Political Science* 19, pp. 33–48

Fukuda-Parr, S. & C. Messineo. 2012. "Human Security: A critical review of the literature." *Centre for Research on Peace and Development (CRPD) Working Paper*, No. 11. Leuven: CRPD

Fukuyama, F. 1989. "The End of History?" In *The National Interest* 26(1), pp. 3–18

Fukuyama, F. 1992. *The End of History and the Last Man*. New York: Free Press

Fukuyama, F. 2011. *The Origins of Political Order from Prehuman Times to the French Revolution*. New York: Farrar, Straus and Giroux

Fukuyama, F. 2014. *Political Order and Political Decay*. New York: Farrar, Straus and Giroux

Fukuyama, F. 2015. "Why Is Democracy Performing So Poorly?" In *Journal of Democracy* 26(1), pp. 11–20

Gerbi, A. 2000. *La Disputa del Nuovo Mondo*, Milano: Adelphi

Gerschenkron, A. 1943. *Bread and Democracy in Germany*. Berkeley: University of California Press

Gerschenkron, A. 1962. *Economic Backwardness in Historical Perspective*. Cambridge, Mass.: Harvard University Press

Giddens, A. 1990. *The Consequences of Modernity*. Stanford: Stanford University Press

Gill, I. & H. Kharas. 2015. "The Middle-Income Trap Turns Ten." *Policy Research Working Paper*, No. WPS 74032015. World Bank Group

Gilman, N. 2003. *Mandarins of the Future: Modernization theory in Cold War America*. Baltimore: Johns Hopkins University Press

Golder, M. 2016. "Far Right Parties in Europe." In *Annual Review of Political Science* 19, pp. 477–498

Göle, N. 2002. "Snapshots of Islamic Modernities." In S. Eisenstadt (ed.), *Multiple Modernities*, pp. 91–117

Gourevitch, P. 1986. *Politics in Hard Times: Comparative responses to international economic crises*. Ithaca, N.Y.: Cornell University Press

Gransow, B. 2006. "Konzeptionen chinesischer Modernisierung: Auf der Suche nach Wohlstand und Stärke." In T. Schwinn (ed.), *Die Vielfalt und Einheit der Moderne*. Wiesbaden: VS Verlag, pp. 151–165

Gries, P. H. 2004. *China's New Nationalism: Pride, politics, and diplomacy*. Berkeley: University of California Press

Gunitsky, S. 2014. "From Shocks to Waves: Hegemonic transitions and democratization in the Twentieth Century." In *International Organization* 68(3), pp. 561–598

Habermas, J. 1985. *Die Neue Unübersichtlichkeit*. Frankfurt/M.: Suhrkamp Verlag

Habermas, J. 2006. *Divided West*. London: Polity Press

Haggard, S. & R. R. Kaufman. 2016. "Democratization During the Third Wave." In *Annual Review of Political Science* 19, pp. 125–144

Hall, P. A. & D. Soskice (eds.) 2001a. *Varieties of Capitalism: The institutional foundations of contemporary advantage*. Oxford: Oxford University Press

Hall, P. A. & D. Soskice. 2001b. "An Introduction to Varieties of Capitalism." In P. Hall & D. Soskice (eds.), *Varieties of Capitalism*, pp. 1–68

Hall, P. A. & R. R. Taylor. 1996. "Political Science and the Three New Institutionalisms." *Political Studies* 44, pp. 936–957

Haq, K. & R. Ponzio. 2008. "Introduction." In K. Haq & R. Ponzio (eds.), *Pioneering the Human Development Revolution*. New Delhi: Oxford University Press

Haq, M. 1995. *Reflections on Human Development*. New York: Oxford University Press

Hartwig, R.P. & C. Wilkinson. 2016. "Cyberrisk: Threat and opportunity." New York: Insurance Information Institute. Available at www.iii.org/white-paper/cyberrisk-threat-and-opportunity-102716/

Harvey, D. 2004. "The 'New' Imperialism: Accumulation by dispossession." *Socialist Register* 40, pp. 63–87

Hellman, J. 1998. "Winners Take All: The politics of partial reform in post-communist transitions." In *World Politics* 50(2), pp. 203–234

Hersant, Y. (ed.). 2000. *Europes. De l'Antiquité au XXe Siècle*. Paris: Laffont

Hettne, B. & A. Inotai, et al. 1999. *Globalism and the New Regionalism*, Vol. 1. Basingstoke: Macmillan

Hobsbawm, E. 1975. *The Age of Empire 1875–1914*. London: Weidenfeld

Hollingsworth, J. R. and Boyer, R. 1997. "Coordination of Economic Actors and Social Systems of Production." In J. R. Hollingsworth & R. Boyer (eds.), *Contemporary Capitalism: The embeddedness of institutions*. Cambridge: Cambridge University Press, pp. 1–47.

Horchani, F. & D. Zolo. 2005. *Mediterraneo: Un dialogo fra le due sponde*. Milano: Jouvence

Huang, Y. 2008. *Capitalism with Chinese Characteristics: Entrepreneurship and the state*. New York: Cambridge University Press

Hung, E. & T.-W. Ngo. 2015. "Parallel Trade in the Pearl River Delta Region." Paper presented at the International Workshop on Cross-border Exchanges and the Shadow Economy: Leiden, December 14–15

Huntington, S. P. 1968. *Political Order in Changing Societies*. New Haven: Yale University Press

Huntington, S. 1996. *The Clash of Civilizations and the Remaking of World Order*. New York: Simon & Schuster

Ikenberry, J. 2017. "The Plot against American Foreign Policy: Can the liberal order survive?" In *Foreign Affairs* 96(3), pp. 2–9

Inglehart, R. 1977. *The Silent Revolution*. Princeton: Princeton University Press

Jain, R. K. & S. Pandey. 2016. "Indian Elites and the EU as a Normative Power." In *Baltic Journal of European Studies* 3(3), pp. 105–126

Jaspers, K. 2011. *The Origin and Goal of History*. London: Routledge Revivals

Johnston, A. I. 2013. "How New and How Assertive is China's New Assertiveness?" In *International Security* 37(4), pp. 7–48

Johnston, A. I. 2016–17. "Is Chinese Nationalism Rising?" *International Security* 41(3), pp. 7–43

Jullien, F. 2016. *Il n'y a pas d'identité culturelle*. Paris: L'Herne

Kant, I. & H. S. Reiss. 1991. *Kant: Political writings*. Cambridge: Cambridge University Press

Kant, I. 1795. "Perpetual Peace: A Philosophical Sketch." Available at http://constitution.org/kant/perpeace.htm

Kant, I. 1966. *Groundwork for the Metaphysics of Morals*, trans. Mary Gregor. Cambridge: Cambridge University Press

Katzenstein, P. 2005. *A World of Regions: Asia and Europe in the American imperium*. Ithaca, N.Y.: Cornell University Press

Katzenstein, P. 2010. *Civilizations in World Politics: Plural and pluralist perspectives*. New York: Routledge

Kaul, I. & D. Blondin. 2016. "Global Public Goods and the United Nations." In J.A. Ocampo (ed.), *Global Governance and Development*. Oxford & New York: Oxford University Press, pp. 32–65

Kaul, I. & P. Conceição (eds.). 2006. *The New Public Finance: Responding to global challenges*. New York: Oxford University Press

Kaul, I. (ed.). 2016. *Global Public Goods*. Northampton, Mass.: Elgar

Kaul, I. 2013. "Meeting Global Challenges: Assessing Governance Readiness." In Hertie-School of Governance (ed.), *The Governance Report 2013*. London: Oxford University Press, pp. 33–58

Kaul, I., D. Blondin, & N. Nahtigal. 2016. "Understanding global public goods: Where we are and where to next." In I. Kaul (ed.), *Global Public Goods*. Cheltenham: Edward Edgar, pp. xiii–xvii

Kaviraj, S. 2002. "Modernity and Politics in India." In S. Eisenstadt (ed.), *Multiple Modernities*, pp. 137–161

Kennedy, P. 1985. *Rise and Fall of Great Powers*. New York: Random House

Keohane, R. 1989. *International Institutions and State Power*. Boulder, CO: Westview Press

Keohane, R. O. 1984. *After Hegemony*. Princeton, N.J.: Princeton University Press

Keohane, R. O. 1986. "Reciprocity in International Relations." In *International Organizations* 40(1), pp.1–27

Kharas, H. & G. Gertz. 2010. "The New Global Middle Class: A cross-over from West to East." In C. Li (ed.), *China's Emerging Middle Class: Beyond economic transformation*. Washington, D.C.: Brookings Institution Press, pp. 32–54

King, A. 2002. "The Emergence of Alternative Modernity in East Asia." In D. Sachsenmaier & J. Riedel (eds.), *Reflections on Multiple Modernities*, pp. 139–152

Kitschelt, H. 1988. "Left-Libertarian Parties: Explaining innovation in competitive party systems." In *World Politics* 40(2), pp. 194–234

Kitschelt, H. 1997. *The Radical Right in Western Europe: A comparative analysis*. Ann Arbor: University of Michigan Press

Kitschelt, H. 2003. "Landscapes of Political Interest Intermediation: Social movements, interest groups, and parties in the early twenty-first century." In P. Ibarra (ed.), *Social Movements and Democracy*. New York: Palgrave, pp. 81–104

Klugman, J., L. Hammer, S. Twigg, T. Hasan, J. McCleary-Sills, & J. Santamaria. 2014. *Voice and Agency: Empowering women and girls for shared prosperity*. Washington, D.C.: World Bank Group

Kocka, J. 2006. "Die Vielfalt der Moderne und die Aushandlung der Universalien." In T. Schwinn (ed.), *Die Vielheit und Einheit der Moderne*. Wiesbaden: VS Verlag, pp. 63–70

Konda Gezi Report. 2014. "Public Perceptions of the Gezi Protesters" (Konda Anket survey, June 5, 2014). Available at Konda Gezi Report.pdf

Kornhauser, W. 1959. *The Politics of Mass Society*. New York: Free Press

Kramer, L. 2014. "Collaboration and 'Diffuse Reciprocity.'" *Stanford Social Innovation Review*, April 25. Available at http://ssir.org/articles/entry/collaboration_and_diffuse_reciprocity

Krasner, S. 1985. *Structural Conflict: The Third World against global liberalism*. Berkeley: University of California Press

Kuo, M.A. 2016. "The End of American World Order: Insights from Amitav Acharya." In *The Diplomat*, November 10. Available at http://thediplomat.com/2016/11/the-end-of-american-world-order/

Kupchan, C.A. 2012. *No One's World: The West, the rising rest and the coming global turn*. New York: Oxford University Press

Lau, D.C. (trans.) 1979b. *Mencius*. Hong Kong, New York: Penguin Books

Lau, D.C. (trans.) 1979a. *Confucius: The Analects*. Hong Kong: The Chinese University of Hong Kong Press

Leatherby, L. 2017. "The rise of antibiotic-resistant infections threatens economies." *Financial Times*. Special Report FT Health: G20, July 7. Available at www.ft.com/content/1a3b06fa-57ff-11e7-80b6-9bfa4c1f83d2?mhq5j=e1/

Lee, K. & S. Li. 2014. "Possibility of a Middle-Income Trap in China: Assessment in terms of the literature on innovation, big business and inequality." In *Frontiers of Economics in China* 9(3), pp. 370–397

Legge, J. (trans.) 1885. *Li Chi* [Book of Rites], vol. 2. Oxford: Oxford University Press; reprinted 1986. Delhi: Motilal Barnarsidass

Legge, J. (trans.) 1960. *The Chinese Classics*, vols. 3 & 5. Hong Kong: Hong Kong University Press

Lerner, D. 1958. *The Passing of Traditional Society: Modernizing the Middle East*. Glencoe, Ill.: Free Press

Liang, Q. 1974. *Xunzi jianzhi*. Hong Kong: Zhonghua shuju

Macklin, R. 2003. "Dignity is a Useless Concept." In *British Medical Journal* 327(7429), pp. 1419–1420

MacLean, S. J., D. R. Black, & T. M. Shaw. [2006] 2016. *A Decade of Human Security: Global governance and new multilateralisms*. London & New York: Routledge

Manners, I. 2002. "Normative Power Europe: A contradiction in terms?" In *JCMS: Journal of Common Market Studies* 40(2), pp. 235–258. Available at doi:10.1111/1468–5965.00353

Manners, I. 2006. "Normative Power Europe Reconsidered: Beyond the crossroads." In *Journal of European Public Policy* 13(2), pp. 182–199. Available at doi:10.1080/13501760500451600

Marx, K. 1992. *Capital, Volume I: A critique of political economy*. London: Penguin Classics

Mauzy, D. K. & R. Milne. 2002. *Singapore: Politics under the People's Action Party*. New York: Routledge

Meyer, T. 2001. *Identity Mania: Fundamentalism and the Politicization of Cultural Differences*. London: Zed

Meyer, T. 2007. "Cultural Differences, Regionalization, and Globalization." In M. Telò (ed.), *European Union and the New Regionalism*. Adlershot, U.K.: Ashgate, pp. 55–75

Meyer, T. 2013. "Cultural Conflicts, Global Governance, and International Institutions." In M. Telò (ed.), *Globalization, Multilateralism, Europe*. Adlershot, U.K.: Ashgate, pp. 287–301

Minkenberg, M. 2000. "The Renewal of the Radical Right: Between modernity and anti-modernity." In *Government and Opposition* 35(2), pp. 170–188

Moffitt, B. 2016. *The Global Rise of Populism: Performance, style, and representation*. Stanford: Stanford University Press

Momin, A. R. No date. "Dr. Mahbub ul Haq (1934–1998)." Available at http://iosminaret.org/vol-5/issue2/profile.php

Moore, B. 1966. *Social Origins of Dictatorship and Democracy*. Boston: Beacon
Moore, M. 1993. *Declining to Learn from the East?* Institute of Development Studies.
Mukherji, R. 2007. *India's Economic Transition: The politics of reforms*. New Delhi: Oxford University Press
Münch, R. 2001. *The Ethics of Modernization*. Lanham, Md.: Rowman & Littlefield Publishers
Nathan, A. 2016. "The Puzzle of the Chinese Middle Class." In *Journal of Democracy* 27(2), pp. 5–19
Naughton, B. 2014. "China's Economy: Complacency, crisis & the challenge of reform." In *Daedalus* 143(2), pp. 14–25
Ngo, T.-W. 2008. "Rent-Seeking and Economic Governance in the Structural Nexus of Corruption in China." In *Crime, Law and Social Change* 49(1), pp. 27–44
Ngo, T.-W. 2009. "The Politics of Rent Production." In T.-W. Ngo & Y. Wu (eds.), *Rent-Seeking in China*. London: Routledge, pp. 1–21
Ngo, T.-W. 2011. "Introduction: Market reform and legacies of the command economy." In T.-W. Ngo (ed.), *Contemporary China Studies: Economy and society*, vol. 1. London: SAGE Publications, pp. xxxv–xxxviii
Ngo, T.-W., C. Yin, & Z. Tang. 2016. "Scalar Restructuring of the Chinese State: The subnational politics of development zones." In *Environment and Planning C (Government and Policy)*, August 9. doi:10.1177/0263774X16661721
Ni, P. 2014. "Seek and You Will Find It: Let Go and You Will Lose It: Exploring a Confucian approach to human dignity." In *Dao: A journal of comparative philosophy* 13(2), pp. 173–198
Nordhaus, W. 2015. "Climate Clubs: Overcoming free-riding in international climate policy." In *American Economic Review* 105(4): 1339–1370
North, D. C., J. J. Wallis, and B. R. Weingast. 2009. *Violence and Social Orders*. Cambridge: Cambridge University Press
Nye, J. 2014. Review of The End of American World Order by A. Acharya. In *International Affairs* 90(5), pp. 1246–1247
Nye, J. 2017. "Will the Liberal Order Survive?" In *Foreign Affairs* (January/February).
O'Donnell, G. 1988. *Bureaucratic Authoritarianism: Argentina, 1966–1973, in comparative perspective*. Berkeley: University of California Press
Osborn, D., J. Cornforth, & F. Ullah. 2014. *National Councils for Sustainable Development: Lessons from the past and present*. SDplanNet Briefing Paper. Ottawa: International Institute for Sustainable Development, & Bangkok: Institute for Global Environmental Strategies
Ozturk, A. 2016. "Examining the Economic Growth and the Middle-Income Trap from the Perspective of the Middle Class." In *International Business Review* 25(3), pp. 726–738
Pagden, A. (ed.). 2002. *The Idea of Europe*. Cambridge: Cambridge University Press
Pagden, A. 2009. *Worlds at War: The 2500 years struggle between East and West*. Oxford: Oxford University Press
Peck, J. & N. Theodore. 2007. "Variegated Capitalism." In *Progress in Human Geography* 31(6), pp. 731–772
Piketty, T. 2014. *Capital in the Twenty-first Century*. Cambridge, MA: The Belknap Press of Harvard University Press
Plato. 1961. *The Collected Dialogues of Plato*, ed. E. Hamilton and H. Cairns. New York: Pantheon
Polanyi, K. 1944. *The Great Transformation*. New York: Farrar & Rinehart
Polanyi, K. 1944. *The Great Transformation: The political and economic origins of our time*. Boston: Beacon Press
Pontusson, J. 2005. "Varieties and Commonalities of Capitalism." In D. Coates (ed.), *Varieties of Capitalism, Varieties of Approaches*. Basingstoke: Palgrave, pp. 163–188
Preyer, G. & M. Sussman. 2015. *Varieties of Multiple Modernities: New research design*. Leiden & Boston: Brill

Przeworski, A., M. Alvarez, J. Cheibub, & F. Limongi. 2000. *Democracy and Development.* Cambridge: Cambridge University Press

Rawls, J. 1999. *The Law of Peoples.* Cambridge, MA: Harvard University Press

Reich, R. 2016. *Saving Capitalism: For the many, not the few.* New York: Vintage Books

Reisen, H. & J. Zattler. 2016. "Shaping the landscape of development finance institutions – World Bank reform as another component of a new world order?" Available at www.kfw-entwicklungsbank.de/PDF/Download-Center/PDF-Dokumente-Development-Research/2015–2001–21-MF-World-Bank_en.pdf/

Robinson, J. & D. Acemoglu. 2005. *Economic Origins of Dictatorship and Democracy.* New York: Cambridge University Press

Rodrik, D. 2017. "Populism and the Economics of Globalization." *NBER Working Paper,* No. 23559. Available at www.nber.org/papers/w23559/

Rostow, W. W. 1960. *The Stages of Economic Growth: A non-communist manifesto.* Cambridge: Cambridge University Press

Rovny, J. 2013. "Where Do Radical Parties Stand? Position blurring in multidimensional competition." In *European Political Science Review* 5(1), pp. 1–26

Roy, O. 2016. *Le Djihad et la Mort.* Paris: Le Seuil

Ruggie, J. 1982. "International Regimes, Transactions, and Change: Embedded liberalism in the postwar economic order." In *International Organization* 36(2), pp. 379–416

Ruggie, J. G. 1992. "International Regimes, Transactions, and Change: Embedded liberalism in the postwar economic order." In S. Krasner (ed.), *International Regimes.* Ithaca, N.Y.: Cornell University Press, pp. 195–232

Sachsenmaier, D. & J. Riedel (eds.). 2002. *Reflections on Multiple Modernities: European, Chinese, and other interpretations.* Leiden & Boston: Brill Publishers

Sayer, D. 1991. *Capitalism and Modernity: An excursus on Marx and Weber.* London: Routledge

Scheve, K. & D. Stasavage. 2012. "Democracy, War, and Wealth: Lessons from two centuries of inheritance taxation." In *American Political Science Review* 106(1), pp. 81–102

Schmidt, V. 2006. "Multiple Modernities or Varieties of Modernity?" In *Current Sociology* 54(1), pp. 77–97

Schumpeter, J. 1934. *The Theory of Economic Development.* Cambridge, Mass.: Harvard University Press

Sen, A. 1999. *Development as Freedom.* New York: Knopf

Sen, A. 2000. "Why Human Security." International Symposium on Human Security. Tokyo, July 28. Available at www.ucipfg.com/Repositorio/MCSH/MCSH-05/BLOQUE-ACADEMICO/Unidad-01/complementarias/3.pdf

Sen, A. 2000b. *Development as Freedom.* New York: Anchor Books

Sen, A. No date, a. Available at hdr.undp.org/en/media/Amartya-Sen-interview-transcript.1.pdf

Sen, A. No date, b. "A 20th Anniversary Hum Dev Discussion with Amartya Sen." Available at www.readbag.com/hdr-undp-en-media-amartya-sen-interview-transcript-1

Shambaugh, D. 2016. "Contemplating China's Future." In *Washington Quarterly* 39(3), pp. 121–130

Shiller, R. J. 2013. "The Risks of the Next Century and Their Management." In I. Palacios-Huerta (ed.), *In 100 Years: Leading economists predict the future.* Cambridge & London: The MIT Press, pp.121–144

Snyder, J. 1991. *Myths of Empire: Domestic politics and international ambition.* Ithaca: Cornell University Press

Snyder, J. 2017. "The Modernization Trap." *Journal of Democracy,* 28(2), pp. 77–91

Stark, D. 1996. "Recombinant Property in East European Capitalism." In *American Journal of Sociology* 101(4), pp. 993–1027

Stiglitz, J. 2003. *Globalization and Its Discontents*. New York: Norton

Stiglitz, J. 2014. "Intellectual Property Rights, the Poll of Knowledge, and Innovation." *NBER Working Paper*, No. 20014. Available at www.nber.org/papers/w20014/

Stiglitz, J. 2017. "Lessons from the Anti-Globalists." Available at www.project-syndicate.org/commentary/macron-fight-against-populism-by-joseph-e–stiglitz-2017–2005?barrier=accessreg/

Stoeckle, K. & M. Rossati. 2018. *Multiple Modernities and Postsecular Societies*. London & New York: Routledge

Streek, W. & K. Yamamura (eds.). 2001. *The Origins of Nonliberal Capitalism: Germany and Japan in comparison*. Ithaca, N.Y.: Cornell University Press

Tadjbakhsh, S. & A. M. Chenoy. 2007. *Human Security: Concepts and implications*. London & New York: Routledge

Tadjbakhsh, S. 2008. "Mahbub ul Haq's Human Security Vision: An Unfettered Dream." In K. Haq & R. Ponzio (eds.), *Pioneering the Hum Dev Revolution: An intellectual biography of Mahbub ul Haq*. Oxford: Oxford University Press

Tan, L. 2017. "War vs. Poor: Police paid per drug killing – Amnesty International." CNN Philippines. http://cnnphilippines.com/news/2017/02/01/war-on-drugs-extrajudicial-killing-Duterte-Amnesty-International.html

Tao, J. 2005. "Relational Resonance with Nature: A Confucian Vision." In I. Lowe & J. Paavola (eds.), *Nature, Justice, and Governance: Environmental values in a globalizing world*. London: Routledge

Teets, J. 2014. *Civil Society under Authoritarianism: The China model*. New York: Cambridge University Press

Telò, M. 2007. "Die Identitäet im 21. Jahrhundert und das fruchtbare Erbe des römischen Europa." In *Leben im Römischen Europa*. Paris: Errance, pp. 22–28

Telò, M. 2016. *Regionalism in Hard Times*. London and New York: Routledge

Telò, M., FawcettL., & PonjaertF., (eds.). 2015. *Interregionalism and the EU*. Farnham: Routledge

Telò, M., C. Ding & X. Zhang (eds.) 2017. *Deepening the EU–China Partnership: Bridging institutional and ideational differences in an unstable world*. Abingdon: Routledge

Thakur, R. 2017. *The United Nations, Peace and Security: From collective security to the responsibility to protect. With a Foreword by Gareth Evans*. 2nd edition. Cambridge: Cambridge University Press

Therborn, G. 1995. *European Modernity and Beyond: The trajectory of European societies, 1945–2000*. London: SAGE

Todorov, T. 1992. *La conquête de l'Amérique: La question de l'autre*. Paris: Le Seuil

Tsai, K. 2007. *Capitalism without Democracy: The private sector in contemporary China*. Ithaca: Cornell University Press

Tu, W. 2002. "Implications of the Rise of 'Confucian' East Asia." In S. Eisenstadt (ed.), *Multiple Modernities*, pp. 115–121

UNESCO Institute for Statistics, 2016, *Statistics of Literacy in China*. Available at http://data.uis.unesco.org

van der Brug, W., M. Fennema, & J. Tillie. 2005. "Why Some Anti-Immigrant Parties Fail and Others Succeed." In *Comparative Political Studies* 38(5), pp. 537–573

van Staen, C. 2016. *La Chine au Prisme des Lumières Françaises*. Bruxelles: Académie Royale de Belgique

von Laue, T. (1987). *The World Revolution of Westernization: The twentieth century in global perspective*. Oxford: Oxford University Press

Wade, R. 2014. "The Art of Power Maintenance." In *Challenge* 56(1), pp. 5–39

Wade, R. H. 2016. "The Role of the State in Escaping the Middle-Income Trap: The case for smart industrial policy." In *METU Studies in Development* 43(1), pp. 21–42

Wagner, P. 2001. "Modernity: one or many?" In J. R. Blau (ed.), *The Blackwell Companion to Sociology*. Oxford: Blackwell Publishing, pp. 30–42

Wakeman Jr., F. 2002. "Chinese Modernity." In D. Sachsenmaier & J. Riedel (eds.), *Reflections on Multiple Modernities*, pp. 153–169

Weber, M. 1930. *The Protestant Ethic and the Spirit of Capitalism*, trans. T. Parsons. London: Allen & Unwin

Weber, M. 1968. "Die drei Typen der legitimen Herrschaft." In J. Winckelmann (ed.), *Max Weber: Soziologische Analysen, Politik*. Stuttgart: Kröner Verlag, pp. 151–166

Weber, S. & B. Jentleson, 2010. *The End of Arrogance: America in the global competition for ideas*. Cambridge, MA: Harvard University Press

Wei, Z. 1978. *Guoyu [jie]*. Shanghai: Shanghai guji chubanshe

Weisbrot, M. & J. Johnston. 2016. *Voting Share Reform at the IMF: Will it make a difference?* Washington, D.C.: Center for Economic and Policy Research. Available at http://cepr.net/images/stories/reports/IMF-voting-shares-2016–2004.pdf/

Weiss, J. C. 2013. "Authoritarian Signaling, Mass Audiences, and Nationalist Protest in China." In *International Organization* 67(1), pp. 1–35

Wilkinson, S. I. 2004. *Votes and Violence: Electoral competition and ethnic riots in India*. Cambridge: Cambridge University Press

Wittrock, B. 2000. "Modernity: One, None, or Many? European origins and modernity as a global condition." In *Daedalus* 129(1), pp. 31–60. Reprinted in *Multiple Modernities*

Wittrock, B. 2002. "Modernity: One, None, or Many? European Origins and Modernity as a Global Condition." In S. Eisenstadt (ed.), *Multiple Modernities*, pp. 31–60

Wolf, M. 2015. *The Shifts and the Shocks: What we've learned – and have still to learn – from the financial crisis*. London: Penguin Books

Wood, E. M. 1997. "Modernity, Postmodernity or Capitalism?" In *Review of International Political Economy* 4(3), pp. 539–560

World Bank Open Data, 2016, *Poverty & Equity Data of China*. Available at http://povertydata.worldbank.org/poverty/country/CHN

World Bank. 1992. "Governance and Development." Washington, D.C.

Yan, L. 2003. *Dongzi Chunqiu fanlu yizhu*, Harbin: Heilongjiang remin chubanshe

Yan, Y. 2010. "The Chinese Path to Individualization." In *British Journal of Sociology* 61(3), pp. 489–512

Yang, B. 1981. *Chunqiu Zuozhuan zhu*, Beijing: Zhonghua shuju

Yang, M.M.-H. 1989. "The Gift Economy and State Power in China." In *Comparative Studies in Society and History* 31(1), pp. 25–54

Yu, K. P. & J. Tao 2012. "Confucianism." In R. Chadwick (ed.), *Encyclopaedia of Applied Ethics*, 2nd edition. San Diego: Academic Press, vol. 1, pp. 578–587

Yu, K. P. 2010. "The Confucian Conception of Harmony." In J. Yao, A. Cheung, M. Painter, & C. Li (eds.), *Governance for Harmony in Asia and Beyond*. London and New York: Routledge, pp. 15–36

Zhang, J. & Peck, J. 2016. "Variegated Capitalism, Chinese Style: Regional models, multi-scalar constructions." In *Regional Studies* 50(1), pp. 52–78

Official Reports

CSIS (Center for Strategic and International Studies). 2014. *Net Losses: Estimating the global cost of cybercrime. Economic impact of cybercrime II.* Prepared in collaboration with McAfee. Washington, D.C.: CSIS. Available at www.mcafee.com/de/resources/reports/rp-economic-impact-cybercrime2.pdf/

ILO (International Labor Office). 2017. *World Employment Social Outlook: Trends for women 2017*. Geneva: ILO.

IPCC (Intergovernmental Panel on Climate Change). 2014a. *Synthesis Report: Summary for Policymakers*. Cambridge & New York: Cambridge University Press

IPCC. 2014b. *Climate Change 2014: Impacts, Adaptation, and Vulnerability. Part A: Global and Sectoral Aspects. Contribution of working group II to the fifth assessment report*. Cambridge & New York: Cambridge University Press

IPCC. 2014c. *Climate Change 2014: Contribution of working group III to the fifth assessment report*. Cambridge & New York: Cambridge University Press

KPMG International. 2014. *Future State 2030: The global megatrends shaping governments*. Available at https://home.kpmg.com/xx/en/home/insights/2015/03/future-state-2030.html

OECD (Organization for Economic Cooperation and Development). 2014. *Boosting Resilience through Innovative Risk Governance*. Paris: OECD

Price Waterhouse Cooper. 2015. *The World in 2050: Will the shift in global economic power continue?* February. Available at www.pwc.com/gx/en/issues/economy/the-world-in-2050.html

UNDP (United Nations Development Program). 1994. *Human Development Report 1994: New dimensions of human security*. New York: Oxford University Press

UNHCR (United Nations High Commissioner for Refugees). 2017. *Global Trends: Forced displacement in 2016*. Geneva: UNHCR

United Nations Development Program. *Human Development Report 2005: International Cooperation at a Crossroads*. New York: United Nations

United Nations Development Program. *Human Development Report 2013. The Rise of the South: Human progress in a diverse world*. New York: United Nations

WBG (World Bank Group). 2013. *World Development Report 2014. Risk and Opportunity; Managing risk for development*. Washington, D.C.: WBG

WBG. 2016. *Progress Report on Mainstreaming Disaster Risk Management in World Bank Group Operations. Document DC2016–0004*. Washington, D.C.: WBG

WEF (World Economic Forum). 2016. *Global Risks 2016*. Geneva: World Economic Forum

WEF. 2015. *Global Risks 2015*. Geneva: World Economic Forum

WHO (World Health Organization). 2015. *Global Action Plan on Antimicrobial Resistance*. Geneva: World Health Organization

WHO. 2007. *The World Health Report 2007 – A safer future: Global health security in the 21st century*. Geneva: World Health Organization

Media

Ashdown, N. 2017. "Turkey moves against dissent." In *U.S. News and World Report*, February 1

The Economist. 2017. "As Venezuela crumbles the regime digs in." January 28. Available at: www.economist.com/news/americas/21715694-nicol-s-maduro-draws-wrong-conclusions-economic-crisis-venezuela-crumbles?fsrc=rss

The Economist. No date. Available at www.economist.com/node/169653

The Guardian Datablog. 2012. "Developing economies to eclipse west by 2060, OECD forecasts." Available at www.guardian.co.uk/global- development/datablog/2012/nov/09/developing-economies-overtake-west-2050-oecd-forecasts

The Irish Times. 2017. January 30. www.irishtimes.com/news/world/europe/hungary-s-orban-eyes-knockout-blow-in-war-with-tycoon-george-soros-1.2955890

INDEX